T0135926

Synchronization of Oscillators and Global Output Regulation for Rigid Body Systems

Der Fakultät 7: Konstruktions-, Produktions- und Fahrzeugtechnik
(Maschinenbau)
und dem Stuttgart Research Centre for Simulation Technology
der Universität Stuttgart zur Erlangung der Würde eines
Doktor-Ingenieurs (Dr.-Ing.) vorgelegte Abhandlung

Vorgelegt von
Gerd Simon Schmidt
aus Rothenburg ob der Tauber

Hauptberichter:	Prof. Dr.-Ing. Frank Allgöwer
Mitberichter:	Prof. Dr. Anders Rantzer
	o. Univ.-Prof. Dipl.-Ing. Dr.techn. Kurt Schlacher

Institut für Systemtheorie und Regelungstechnik der Universität Stuttgart

2013

Bibliografische Information der Deutschen Nationalbibliothek

Die Deutsche Nationalbibliothek verzeichnet diese Publikation in der
Deutschen Nationalbibliografie; detaillierte bibliografische Daten sind
im Internet über http://dnb.d-nb.de abrufbar.

ISBN 978-3-8325-3790-6

Logos Verlag Berlin GmbH
Comeniushof, Gubener Str. 47,
10243 Berlin
Tel.: +49 (0)30 42 85 10 90
Fax: +49 (0)30 42 85 10 92
INTERNET: http://www.logos-verlag.de

Acknowledgements

The results presented in this thesis are part of the outcome of my time as a research and teaching assistant at the Institute for Systems Theory and Automatic Control (IST), University of Stuttgart. Throughout this time, many people influenced this thesis and my life directly or indirectly. Since the only written record of my interactions are my publications, I would like to take this chance to express my gratitude to the people around me. Any person I mention here takes credit for the thesis being what it is now.

I want to thank Prof. Dr.-Ing. Frank Allgöwer for the opportunity to work, learn and teach at the IST and to be part of an outstanding research environment. I am extremely grateful for this opportunity, which has been the basis for my personal and my scientific development during the last years. Furthermore I want to thank him for his commitment to the Stuttgart Research Centre for Simulation Technology, which offered a great number of additional possibilities of interdisciplinary and international interaction. Even though his schedule is very busy, he always had the time to give support and advice when needed.

I also want to express my thanks to Prof. Dr.-Ing. Christian Ebenbauer for the effect he had on my thinking and my work. I am very grateful for his exclusively positive influence on my scientific output, especially on the thesis, through our collaborations and countless hours of detailed discussion about various research problems. He was always a very patient first reader of my writing and provided consistently constructive and insightful criticism. In addition to that, he also offered invaluable personal advice and was one of the people at the institute with whom I had very interesting non-academic discussions.

Furthermore, I want to express my deepest thanks to Prof. Dr.-Ing. Arnold Kistner, Prof. Anders Rantzer and o. Univ.-Prof. Dipl.-Ing. Dr.techn. Kurt Schlacher, for the interest in my work, for the insightful and interesting comments on my work and for being members of my doctoral examination commitee.

Many thanks go also to Prof. Jeff Moehlis for the possibility to spend more than three months with his group at the University of California in Santa Barbara. My time there was invaluable for my academic and my personal development. I am very grateful for the successful research collaboration and the interaction with Jeff and his group.

An important part of my academic life were also my teachers, whom I want to thank for their invaluable commitment and for the contagious enthusiam for their subjects and the academic world in general. In that role, I especially want to mention Prof. Dr.-Ing. Frank Allgöwer, Prof. em. Dr. Gerd Blind, Prof. Dr.-Ing. Christian Ebenbauer, Prof. Dr. Richard Dipper, Prof. Dr.-Ing. Arnold Kistner, Prof. Dr.-Ing. Herbert Wehlan, Prof. Dr. Annette Werner and Prof. Dr.-Ing Michael Zeitz.

I also want to thank my colleagues for being available for vivid scientific discussions. Some of these resulted in very interesting research collaborations, for which I want to thank Mathias Bürger, Simon Michalowsky, Ulrich Münz, Georg Seyboth, Jan-Maximilian Montenbruck and Peter Wieland.

Not mentioning that my time at the institute has been especially pleasant because of the interaction with some colleagues would be the greatest omission. Therefore I want to express my special thanks to Christian Breindl, Hans-Bernd Dürr, Eva-Maria Geissen, Dirke Imig, Andrei Kramer, Beate Spinner, Patrick Weber and Peter Wieland for making my time at the institute really enjoyable by being available for all sorts of discussions or for spare time activities. Futher thanks go to all my colleagues and guests at the institute for every academic or non-academic interaction.

Finally, I also want to express my great gratitude to my parents and my siblings for their continuous support over many years and for all the personal advice they offered.

Contents

List of Figures

Abstract

The objective of this thesis is to provide non-local or global analysis and synthesis results for synchronization in oscillator networks and for output regulation problems for rigid body systems.

Synchronization problems in oscillator networks are concerned with the analysis of the stability or the attractivity of periodic solutions in systems consisting of a possibly large number of coupled oscillators. Due to the non-trivial steady-state behavior, the non-local or global analysis of oscillator networks is important for the investigation of synchronization. Here, we consider synchronization in two different model classes, namely in so-called Kuramoto models and in oscillator networks consisting of identical oscillators where we know a Lyapunov function for the periodic orbit of the single oscillator. For the Kuramoto models we investigate the impact of delays on two synchronization effects and provide new conditions that guarantee synchronization. For the oscillator networks consisting of identical oscillators where we know a Lyapunov function for the periodic orbit of the single oscillator, we give a new characterization of synchronization and give conditions that guarantee synchronization. The provided sufficient conditions for synchronization are especially useful for the design of couplings for oscillators to achieve synchronization in the associated oscillator network.

Output regulation problems are concerned with the synthesis of controllers that achieve asymptotic disturbance rejection and asymptotic tracking. Here, we consider a class of output regulation problems which includes attitude tracking control problems for rigid body systems. From an application perspective it is desirable that the controller achieves the control objective for all configurations, in other words that the associated output regulation problem is solved globally. However, due to the geometry of the state space of rigid body systems, continuous vector fields necessarily have multiple equilibria in the presence of an asymptotically stable equilibrium. The synthesis of a controller to solve the associated output regulation problem includes a stabilization problem which contains the stabilization of a point on the state space of the rigid body system. Hence, the design of smooth control laws necessarily results in closed loop systems with multiple equilibria, which poses a challenge for established synthesis methods. Here, we propose a new two-step control design procedure that provides a solution for the considered class of output regulation problems and that results in a closed loop system where any solution converges to one of the finitely many isolated equilibria, where the desired equilibrium is the only asymptotically stable equilibrium and where the other equilibria are unstable. The two steps are a state feedback design and an observer design, and the output feedback controller is a certainty equivalence implementation of the state feedback. Since the design of the state feedback and the observer are independent, the proposed design method establishes a new separation principle for the considered problem class.

Deutsche Kurzzusammenfassung

Die Modellierung vieler Prozesse in ingenieurtechnischen Anwendungen resultiert in einem nichtlinearen Differentialgleichungsmodell. Wegen der zunehmenden Anforderungen an die Zuverlässigkeit und die Effizienz der Prozesse spielen nicht-lokale beziehungsweise globale dynamische Eigenschaften der Modelle eine zunehmende Rolle in der Analyse der Prozesse und beim modellbasierten Reglerentwurf. Eine Herausforderung für die nicht-lokale beziehungsweise globale Analyse und für entsprechende Reglerentwurfsverfahren ist in diesem Fall die Existenz von mehreren Ruhelagen oder das Auftreten einer nicht-trivialen stationären Dynamik, wie z.b. einer periodischen Lösung. Diese Arbeit beschäftigt sich mit zwei Problemklassen, bei denen die vorher genannten Effekte inhärent vorhanden sind und bei denen die nicht-lokale beziehungsweise globale Sichtweise eine wichtige Rolle spielt, nämlich der Synchronisation von Oszillatoren und der Ausgangsregelung von Starrkörpersystemen. Für beide Problemklassen liegen eine Vielzahl von Ergebnissen vor. Ungeachtet dessen gibt es für beide Problemklassen noch viele unbearbeitete Fragestellungen. Das übergeordnete Ziel dieser Arbeit ist es nicht-lokale beziehungsweise globale Analyse- und Synthese-Fragestellungen zu identifizieren und zu bearbeiten. Dabei wird eine systemtheoretische Herangehensweise verwendet. Das heißt, Erkenntnisse über die Eigenschaften der betrachteten System sollen möglichst mit Hilfe der Eigenschaften von Teilsystemen erzielt werden. Die Teilsystem sollen dabei so gewählt werden, dass sich Erkenntnisse über die Eigenschaften der Teilsysteme verhältnismässig einfach gewinnen lassen. Aus einer theoretischen Perspektive sind die Ergebnisse dieser Arbeit der Stabilitätstheorie für nichtlineare Systeme zuzuordnen.

Mit Synchronisation wird ein Phänomen bezeichnet, welches häufig in Systemen beobachtet wird, die aus gekoppelten Oszillatoren bestehen. Vereinfacht gesprochen bezeichnet man mit Synchronisation das Phänomen, bei dem die Kopplung die Oszillatoren asymptotisch zu einer identischen oszillatorischen Dynamik führt. Ein typisches Beispiel für Synchronisation von Oszillatoren in einem technischem System ist die Synchronisation von Generatoren in einem elektrischen Netzwerk. Im einfachsten Fall handelt es sich um eine Menge von Generatoren, die durch Leitungen und Lasten gekoppelt sind. Um den Betrieb zu ermöglichen, ist es dabei notwendig, dass die von den Generatoren erzeugte Wechselspannung mit gleicher Frequenz und gleicher Phase, also synchron, in das Netz eingespeist wird. Weiterhin soll der synchrone Betrieb robust gegenüber Laständerungen und Leitungsausfällen sein. Um diese Effekte theoretisch zu untersuchen, werden die Generatoren als Oszillatoren und die Leitungen und Lasten als Kopplung modelliert. Die Modelle, die sich dabei ergeben, sind allerdings unabhängig von dieser speziellen Anwendung. Man erhält ähnliche Modelle, wenn man durch Diffusion gekoppelte chemische Oszillatoren betrachtet oder Rotoren modelliert, die auf einer elastisch gelagerten Platte montiert sind. Folglich werden die entsprechenden Modellklassen häufig unabhängig von der Anwendung betrachtet. In der vorliegenden Arbeit wird Synchronisation von Oszillatoren sowohl direkt für Modelle gekoppelter Oszillatoren, sogenannter Oszillatornetzwerke, als auch für Kuramoto-Modelle untersucht.

Kuramoto-Modelle gehören zur Klasse der Phasenmodelle, die ein vereinfachtes und niedrigdimensionales Modell für die Dynamik eines Oszillatornetzwerkes sind. Solch eine Beschreibung ist möglich, wenn die periodischen Orbits der Oszillatoren starke Stabilitätseigenschaften haben und die Anfangsbedingungen der Oszillatoren in der Nähe ihrer periodischen Orbits liegen. Es handelt sich also um eine lokale Beschreibung des Oszillatornetzwerkes. Synchronisation im Oszillatornetzwerk kann dann durch die Stabilitätsanalyse spezieller Lösungen im Phasenmodell untersucht werden. Ein Problem ist dabei die typischerweise nichttriviale Dynamik der Phasenmodelle. In dieser Arbeit wird der Einfluß von Totzeiten in den Kopplungen auf die Synchronisation im Kuramoto-Modell untersucht. Man kann erwarten, dass eine stärkere Kopplung zwischen den einzelnen Oszillatoren dazu führt, dass der Einfluß der Totzeiten kompensiert wird. In dieser Arbeit werden entsprechende Bedingungen angegeben, mit deren Hilfe sich die Synchronisation in diesen Modellen nachweisen lässt. Diese Untersuchungen verbessern auch die Erkenntnisse über die entsprechenden Modelle ohne Totzeiten.

In den vorliegenden direkten Untersuchungen der Oszillatornetzwerke wird angenommen, dass die Stabilität des periodischen Orbits mittels einer Lyapunovfunktion nachgewiesen werden kann. Bearbeitet werden in diesem Fall Fragestellungen, die nicht durch die Analyse mit Phasenmodellen abgedeckt werden, d.h. Bedingungen, die Synchronisation garantieren, wenn die Anfangsbedingungen nicht in der Nähe der periodischen Orbits liegen. Solche Bedingungen sind wichtig, wenn man Kopplungsfunktionen für Oszillatoren entwirft und Synchronisation für möglichst viele Anfangskonfigurationen garantieren möchte. Kopplungsfunktionen werden häufig mit dem Ziel entworfen, dass der Fehler zwischen den Zuständen der Oszillatoren im Oszillatornetzwerk gegen Null konvergiert. Das heißt, die Kopplungsfunktionen werden mit dem Ziel entworfen, die Attraktivität und die Invarianz der Mannigfaltigkeit zu garantieren, auf welcher der Fehler zwischen den Zuständen verschwindet. Man nennt diese Mannigfaltigkeit auch Synchronisationsmannigfaltigkeit. Eine wichtige Frage in diesem Kontext ist, ob die Attraktivität und Invarianz der Synchronisationsmannigfaltigkeit auch die Synchronisation der Oszillatoren garantiert. Die Schwierigkeit in diesem Fall ist die nichttriviale Dynamik des Oszillatornetzwerkes auf der Synchronisationsmannigfaltigkeit. In dieser Arbeit werden einige wichtige Aspekte der auftretenden Probleme im Detail diskutiert. Des Weiteren werden hinreichende Bedingungen angegeben, die Synchronisation für die Oszillatoren garantieren, bei denen eine Lyapunovfunktion für die periodischen Orbits existiert. Die diskutierten Ansätze stellen eine neuartige Sichtweise auf das Problem der Oszillatorsynchronisation dar.

Im Gegensatz zu den Synchronisationsproblemen, die in der Hauptsache Analyseprobleme sind, ist das Ausgangsregelungsproblem für Starrkörpersysteme ein Entwurfsproblem. Beim Ausgangsregelungsproblem werden Regelstrecken betrachtet, die von einer Familie von Störungen beeinträchtigt werden. Ziel ist es, dass eine Funktion der Zustände des Systems, also ein Systemausgang, asymptotisch einer Familie von Referenzen folgt und gleichzeitig der Einfluß der Störungen asymptotisch unterdrückt wird. Betrachtet man die Differenz zwischen dem Ausgang und den Referenzen als neuen Ausgang, den sogenannten *zu kontrollierenden Ausgang*, so ist das Ausgangsregelungsproblem äquivalent dazu, den zu kontrollierenden Ausgang für alle Störungen aus der Familie der Störungen und alle Referenzen aus der Familie der Referenzen asymptotisch zu einer Konstante zu regeln. In der Theorie hat es sich dabei etabliert anzunehmen, dass sowohl die Referenzen, als auch die Störungen Lösungen einer bekannten Differentialgleichung sind. Diese Differentialgleichung ist das sogenannte *Exosystem*.

In dieser Arbeit wird eine Klasse von nichtlinearen Ausgangsregelungsproblemen betrachtet, in welchen die globale Sichtweise auf die Systemdynamik eine wichtige Rolle spielt. Das heißt, in dieser Arbeit werden Ausgangsregelungsprobleme für Starrkörpersysteme betrachtet. Ein Starrkörpersystem ist ein mechanisches System bestehend aus Massepunkten, welche durch die holonome Zwangsbedingung eingeschränkt sind, dass der Abstand zwischen jedem Paar von Massepunkten konstant bleibt. Näherungsweise ist diese Annahme für viele mechanische Systeme oder Teilsysteme erfüllt. Deswegen stellen Starrkörpersysteme eine wichtige Klasse mechanischer Modelle dar. Für viele Anwendungen, zum Beispiel in der Robotik und allgemeiner für Mehrkörpersysteme, ist es wichtig dass die verwendeten Regler für alle Konfigurationen des mechanischen Systems, also global, anwendbar sind. Eine besondere Eigenschaft eines starren Körpers ist der Zustandsraum für dessen rotatorische Bewegung. Dieser Zustandsraum ist durch die Menge der dreidimensionalen Rotationsmatrizen $SO(3) = \{\Theta \in \mathbb{R}^{3\times3} | \Theta^{-1} = \Theta^{\mathsf{T}}, \det(\Theta) = 1\}$ gegeben. Die nichttriviale Geometrie von $SO(3)$ stellt eine Schwierigkeit für die Stabilisierung einer Ruhelage mittels eines glatten Rückführgesetzes dar, da jedes stetige Vektorfeld auf $SO(3)$ mit einer asymptotisch stabilen Ruhelage notwendigerweise weitere Ruhelagen hat. Somit kann keine globale asymptotische Stabilität der gewünschten Ruhelage mittels glatter Rückführgesetze gewährleistet werden. Stattdessen kann man in diesen Situationen lediglich erreichen, dass ein glattes Rückführgesetz die Stabilität der gewünschten Ruhelage für alle Anfangsbedingungen bis auf eine vernachlässigbar kleine Menge garantiert. Man spricht dann von fast globaler Stabilität. Eine schwächere, aber leichter erzielbare Eigenschaft wäre die garantierte Konvergenz jeder Lösung zu einer von endlich vielen isolierten invarianten Mengen unter der Bedingung, dass lediglich eine Menge asymptotisch stabil ist und die anderen Mengen instabil sind. Da in jedem Ausgangsregelungsproblem auch ein Stabilisierungsproblem enthalten ist und in dieser Arbeit glatte Rückführgesetze verwendet werden sollen, führt jedes global anwendbare Rückführgesetz auf einen geschlossenen Kreis, in welchem mehrere Ruhelagen existieren. Die gleichen Einschränkungen gelten analog für Regelungssysteme deren Zustandsraum $SO(n)$ oder $SE(n) = \mathbb{R}^n \times SO(n)$ ist, wobei $SO(n)$ die Menge $\{\Theta \in \mathbb{R}^{n\times n} | \Theta^{-1} = \Theta^{\mathsf{T}}, \det(\Theta) = 1\}$ bezeichnet. Die gerade beschriebenen, probleminhärenten Eigenschaften stellen ein Hindernis für den Einsatz von existierenden Entwurfsmethoden dar. Dies gilt im Besonderen, falls nicht die gesamte Zustandsinformation der Strecke und des Exosystems durch Messungen gewonnen werden kann.

In der vorliegenden Arbeit wird ein neuartiges beobachterbasiertes Entwurfsverfahren zur Synthese von Ausgangsrückführungen vorgestellt, die eine Lösung für die entsprechenden Ausgangsregelungsprobleme darstellen. Die Besonderheiten des Verfahrens sind, dass wir global anwendbare Ausgangsrückführungen entwerfen können, die die oben angefürten Stabilitätseigenschaften der gewünschten Ruhelage garantieren. Eine weitere Besonderheit ist der zweistufige Charakter des Entwurfsverfahrens. Das heißt, man entwirft zuerst eine Zustandsrückführung, die fast globale Stabilität der gewünschten Ruhelage garantiert und im Anschluß einen Beobachter. Das Ausgangsrückführgesetz erhält man dann, indem man die nicht messbaren Zustände durch die entsprechenden beobachteten Zustände ersetzt. Die wichtigste Eigenschaft der vorgestellten Lösung ist nun, dass die Konvergenzeigenschaften der Strecke mit der Ausgangsrückführung mit Hilfe der Konvergenzeigenschaften der Strecke mit der Zustandsrückführung und den Konvergenzeigenschaften des Beobachters untersucht werden können. Das heißt, die Existenz einer sogenannten Höhenfunktion (*height function*, siehe auch Abschnitt 1.2.2) mit gewissen Eigenschaften für das System mit Zustandsrückführung zusammen mit einem konvergenten Beobachter sind hinreichend, um die Konvergenz im geschlossenen Kreis mit der Ausgangsrückführung zu garantieren. Die Konvergenz wird garantiert für alle konvergenten Beobachter zusammen mit allen Zustandsrückführungen, für welche die Eigenschaften der Höhenfunktion erhalten bleiben. Aus diesem Grund beinhaltet das Resultat ein Separationsprinzip für die betrachtete Klasse von nichtlinearen Problemen. Das heißt, das Entwurfsverfahren ähnelt dem bekannten Ausgangsrückführungsresultat aus der linearen Systemtheorie, in welchem die Stabilität des geschlossenen Kreises garantiert ist, obwohl die Pole des Beobachters als auch die Pole der Zustandsrückführung unabhängig gewählt werden. Die vorgestellte Entwurfsmethode und das Separationsresultat sind nicht auf das Starrkörperproblem beschränkt und können in analogen Situationen ebenso eingesetzt werden.

Forschungsbeiträge und Gliederung der Arbeit

Die Arbeit ist thematisch gegliedert, d.h. die Beiträge zum Thema *Synchronisation von Oszillatoren* werden im Kapitel 2 präsentiert und die Beiträge zum Thema *Ausgangsregelung für Starrkörpersysteme* werden im Kapitel 3 präsentiert. Jedes Kapitel enthält jeweils zwei Beiträge zum jeweiligen Thema. Die Kapitel sind in sich wie folgt strukturiert; zuerst wird eine allgemeine Einleitung zum jeweiligen Thema präsentiert, in welcher die notwendigen Grundbegriffe erklärt und die wichtigste Hintergrundliteratur diskutiert wird. Im Anschluss daran wird jeder Beitrag in einem eigenen Abschnitt dargestellt. Diese Abschnitte beginnen mit einer Problemstellung, gefolgt von beitragsspezifischer Hintergrundliteratur. Anschließend werden die Resultate zum jeweiligen Beitrag präsentiert. Jeder Abschnitt wird mit einer kurzen Zusammenfassung und jedes Kapitel mit einer kurzen Diskussion beendet. Die Beiträge der Arbeit werden abschließend in Kapitel 4 zusammengefasst und diskutiert.

Beiträge in Kapitel 2 — Synchronization of oscillators

Das Kapitel 2 beschäftigt sich mit Synchronisationsproblemen in Oszillatornetzwerken.

Im Abschnitt 2.1 werden die grundlegenden Begriffe eingeführt, die für die Diskussion der Synchronisationsprobleme in Oszillatornetzwerken notwendig sind. Insbesondere werden die Eigenschaften der *orbitalen Stabilität* und der *asymptotischen Phase* der periodischen Orbits von Oszillatoren diskutiert. Es wird definiert, was unter einem Oszillatornetzwerk, einem Phasenmodell eines Oszillators und eines Oszillatornetzwerkes zu verstehen ist. Des Weiteren werden die Synchronisationseffekte in Oszillatornetzwerken erklärt, die in dieser Arbeit betrachtet werden. Dabei wird auch kurz der Zusammenhang zwischen Synchronisationsproblemen in Oszillatornetzwerken und der Attraktivität spezieller Lösungen im Phasenmodell erörtert.

Im Abschnitt 2.2 wird die Synchronisation in Kuramoto-Modellen mit Totzeiten diskutiert. Dieser Abschnitt basiert auf Schmidt et al. [125; 126]. Es werden neuartige notwendige und hinreichende Bedingungen für zwei Synchronisationseffekte in Kuramoto-Modellen mit Totzeiten diskutiert. Bei den hinreichenden Bedingungen wird zwischen den Fällen unterschieden, in welchen jeder Oszillator mit jedem anderen gekoppelt ist und solchen, in welchen dies nicht der Fall ist. Im ersten Fall ergeben sich einfach überprüfbare Bedingungen, die Synchronisation garantieren, im zweiten Fall hängen die Bedingungen von der Lösung einer nichtlinearen Gleichung ab.

Im Abschnitt 2.3 wird die Synchronisation in Netzwerken von identischen Oszillatoren untersucht, bei denen eine Lyapunovfunktion für die periodischen Orbits der ungekoppelten Oszillatoren existiert. Dieser Abschnitt basiert auf Schmidt et al. [119]. Der erste Beitrag dieses Abschnittes ist eine Diskussion, die erklärt, welche Schwierigkeiten aus der Sicht der nichtlinearen Dynamik bei der Synchronisation von Oszillatoren auftreten können. Daraufhin wird eine neuartige Charakterisierung von Synchronisation in Oszillatoren präsentiert, die aus zwei Eigenschaften besteht. Im Folgenden wird nachgewiesen, dass die Annahme einer Lyapunovfunktion für die periodischen Orbits der Oszillatoren dazu verwendet werden kann, eine der beiden Eigenschaften zu erfüllen. Des Weiteren, wird eine Bedingung an den synchronen Orbit angegeben, mittels derer die zweite Eigenschaft garantiert wird. Das heißt, es werden Bedingungen angegeben, die hinreichend für die Synchronisation in Oszillatornetzwerken sind.

Beiträge in Kapitel 3 — Global output regulation and the regulation of rigid bodies

Das Kapitel 3 beschäftigt sich mit Ausgangsregelungsproblemen für Systeme auf $SE(n)$ und für Starrkörpersysteme.

Im Abschnitt 3.1 werden kurz die Grundlagen der Ausgangsregelung erklärt. Weiterhin wird die zentrale Schwierigkeit beim Entwurf einer Ausgangsrückführung erläutert.

Im Abschnitt 3.2 wird eine Klasse von Ausgangsregelungsproblemen auf $SE(n)$ untersucht. Die ersten Ergebnisse dieses Abschnittes wurden in Schmidt et al. [120] präsentiert. Teile dieses Abschnittes werden in Schmidt et al. [123] veröffentlicht. Der erste Beitrag dieses Abschnittes ist die Präsentation eines beobachterbasierten Ansatzes zum Entwurf einer Ausgangsrückführung, welche das Ausgangsregelungsproblem für die betrachtete Systemklasse löst. Die Lösung ist global und garantiert die Konvergenz alle Lösungen zu einer von endlich vielen isolierten invarianten Mengen, von denen genau eine asymptotisch stabil ist und die anderen sind instabil, sowie die Beschränktheit der Lösungen im geschlossenen Kreis für alle Anfangsbedingungen. Das Entwurfsverfahren ist zweigeteilt. Ein Teil ist der Entwurf einer Zustandsrückführung, die sowohl die Zustande der Strecke als auch die Zustände des Exosystems verwendet, um die gewünschten Konvergenzeigenschaften im geschlossenen Kreis mit der Zustandsrückführung zu erreichen. Hier bedeutet das die Konvergenz des zu kontrollierenden Ausgangs zum gewünschten Wert für fast alle Anfangsbedingungen. Der zweite Teil ist der Entwurf eines Beobachters für die nicht gemessenen Zustände. Im vorliegenden Fall sind dies Zustände des Exosystems. Die Ausgangsrückführung besteht aus dem Beobachter und der Zustandsrückführung, in welcher die nicht gemessenen Zustände durch die entsprechenden Beobachterzustände ersetzt werden. Trotz des unabhängigen Entwurfs der beiden Teile kann man im geschlossenen Kreis mit der Ausgangsrückführung die oben gennanten Konvergenzeigenschaften garantieren. Beim Konvergenzbeweis verwendet man eine sogenannte Höhenfunktion. Der zweite Beitrag dieses Abschnittes ist ein Separationsprinzip in Analogie zum linearen Fall. Dies ist wie folgt begründet: Der Konvergenzbeweis gilt für alle Zustandsrückführungen, bei denen die Eigenschaften der Höhenfunktion unverändert bleiben, und alle konvergenten Beobachter. Das heißt man kann die Zustandsrückführung und den Beobachter austauschen, ohne die Konvergenzeigenschaften zu verlieren. Weil der Entwurf der beiden Teile unabhängig ist, kann man im vorliegenden Fall von einem Separationsprinzip sprechen.

Im Abschnitt 3.3 wird eine Klasse von Ausgangsregelungsproblemen für Starrkörpersysteme untersucht. Dieser Abschnitt basiert auf Schmidt et al. [122] und Schmidt et al. [124]. Der erste Beitrag dieses Abschnittes ist die Erweiterung des Entwurfsverfahrens aus Abschnitt 3.3, um eine Ausgangsrückführung zu entwerfen, die eine Lösung für die betrachtete Klasse von Ausgangsregelungsproblemen für Starrkörpersysteme darstellt. Dabei muss der Entwurf der Zustandsrückführung und auch der Beobachterentwurf entsprechend angepasst werden. Der Beweis der Konvergenz verläuft analog zum Konvergenzbeweis im Abschnitt 3.2. Der zweite Beitrag des Abschnittes ist die Implementierung der entwickelten Theorie für ein Anwendungsszenario, in welchem die Lage eines Satelliten geregelt wird. Das betrachtete Satellitenmodell besteht aus drei Starrkörpern. Einem Hauptkörper, dessen Lage geregelt werden soll und zwei symmetrischen Solarmodulen die einen rotatorischen Freiheitsgrad haben.

Im Abschnitt 3.4 werden die Konvergenzeigenschaften einer Differentialgleichung auf $SO(n)$ betrachtet. Dieser Abschnitt basiert auf Schmidt et al. [121]. Die betrachtete Differentialgleichung taucht als Fehlerdynamik oder Teil der Fehlerdynamik in Abschnitt 3.2 und Abschnitt 3.3 auf. Es wird gezeigt, dass es sich bei der betrachteten Fehlerdynamik um einen Gradientenfluß handelt. Der erste Beitrag des Abschnittes ist die Untersuchung der Eigenschaften der Funktion. Dabei stellt sich heraus, dass die Funktion eine sogenannte Morse-Bott-Funktion ist. Daraus folgt, dass die Lösungen des Gradientenflusses für alle Anfangsbedingungen zu einem kritischen Punkt konvergieren. Der zweite Beitrag des Abschnittes ist der Nachweis der Konvergenz der Lösungen für fast alle Anfangsbedingungen zu genau einem kritischen Punkt, der der gewünschten Ruhelage entspricht.

Beiträge in Kapitel 4 — Conclusions

Im Schlusskapitel dieser Arbeit, also im Kapitel 4, wird eine Abschlussbetrachtung der Arbeit vorgenommen. Des Weiteren, werden zukünftige Forschungsmöglichkeiten ausgehend von den Beiträgen in Kapitel 2 und Kapitel 3 beschrieben.

Insbesondere wird die Rolle des Exosystem für die Anwendbarkeit der Theorie der Ausgangsregelung in praktischen Problemen hervorgehoben. Ferner wird vorgeschlagen, wie die neuartige Charakterisierung der Oszillatorsynchronisation aus dem Abschnitt 2.3 für weitere Untersuchungen von Effekten in Oszillatornetzwerken verwendet werden kann. Im Falle der Synchronisationsuntersuchungen mittels Kuramoto-Modellen im speziellen oder Phasenmodellen im allgemeinen, wird vor allem auf die offenen Probleme im totzeitfreien Fall eingegangen, welche vor dem totzeitbehafteten Fall gelöst werden sollten.

1

Introduction

The non-local or global analysis and control of nonlinear dynamical systems is of great importance for many engineering applications. One challenge for global analysis and control of nonlinear dynamical systems is the presence of non-trivial steady-state behavior, such as periodic orbits, and of multiple equilibria. In this work, we consider two specific problem classes where non-trivial steady-state behavior or the presence of multiple equilibria is inherent and where the non-local or global point of view plays an important role. More precisely, we consider synchronization problems for oscillator networks and global output regulation problems for rigid body systems. Both problem classes have been the subject of a large research effort and still offer many challenges. The overall objective of this thesis is to provide non-local or global analysis and synthesis results for these problem classes. We aim for a system theoretic approach to obtain these results. This means in the current context to determine the properties of the complete system by utilizing known or relatively simple to analyze properties of a subsystem or of several subsystems. From the theoretical perspective, the contributions of this work are in the area of stability theory for nonlinear dynamical systems.

Synchronization problems concern an effect which can be observed in a system of coupled oscillators, a so-called oscillator network. Roughly speaking, synchronization in an oscillator network means that the coupling between the oscillators leads to asymptotically common oscillatory dynamics for the individual oscillators. A typical example of synchronization in an engineering system is the synchronization of generators in an electrical power network. This example concerns electrical generators coupled in the simplest case through a system consisting of transmission lines and loads. A necessary condition for the operation of the power network is that the generators run synchronously and that they maintain the synchrony under disturbances, such as the loss of transmission lines or changes of loads. For theoretical investigations of synchronization in power networks, the electrical network is modeled as an oscillator network where the oscillators represent the generators. However, the models utilized to investigate synchronization for electrical power networks are independent of this specific application example. Similar models are also used to investigate for example synchronization of chemical oscillators coupled by diffusion or the synchronization of rotors mounted on a common support. As a consequence, there is an interest in theoretical studies of oscillator synchronization in these classes of oscillator networks.

We consider in this work oscillator synchronization in two different system classes. First, we consider synchronization problems for the Kuramoto model. The Kuramoto model belongs to the class of phase models, which are simplified lower-dimensional models for a class of oscillator networks. The phase model approach is possible if the periodic orbits of the oscillators have strong stability properties and all systems are initialized in the vicinity of the periodic orbits. Synchronization of the oscillators in the oscillator network can then be investigated by analyzing the stability or attractivity of specific solutions of the phase model. One challenge for this analysis are the typically non-trivial dynamics of the phase models. The problem we investigate here is the influence of delays in the Kuramoto model on the synchronization behavior. The investigation of delays in the Kuramoto model is justified if the delays in the oscillator network are large. One can expect that the delays do not affect synchronization if the coupling is strong enough to compensate the effect of the delays. Indeed, we can give novel conditions which ensure synchronization in the Kuramoto model with delays. In addition, our considerations also improve the analysis of synchronization in the Kuramoto model without delays. The second system class we consider are oscillator networks with a Lyapunov stability property for the periodic orbit of the individual oscillators. We are interested in conditions that are not covered by a phase model analysis, especially conditions for the coupling functions which guarantee synchronization if the initial conditions of the system can range over a large subset of the state space. Conditions that guarantee synchronization in such a situation are of interest for the design of synchronizing couplings. A common goal for the design of synchronizing couplings is to drive the error between the states of the oscillators to zero. In other words one tries to achieve attractivity and invariance of the zero error manifold. The question is whether the approach of achieving attractivity and invariance of the zero error manifold, also called synchronization manifold, is sufficient to guarantee synchronization. The challenge in this context are the dynamics of the oscillator network on the synchronization manifold, which are typically non-trivial. Here, we present a detailed discussion of some of the possible difficulties that appear in that context. Furthermore, we present novel sufficient conditions that guarantee synchronization for the considered class of oscillators. Our analysis offers a new approach to analyze oscillator networks based on the properties of the individual oscillators and a property of the synchronous solution.

In contrast to the analysis problems concerning synchronization of oscillator networks, the output regulation problems for rigid body systems are concerned with synthesis. The theory of output regulation studies control systems which are affected by a family of disturbances. The goal is to design a controller such that an output of the control system asymptotically tracks a family of references while asymptotically rejecting the family of disturbances. If we consider the error between the references and the output that should track the references as a new output, called the regulated output, then the output regulation problem is equivalent to the design of a controller which achieves asymptotic convergence of the regulated output to a constant value for all references and all disturbances. An integral and well established assumption to solve output regulation problems is that the references and disturbances are the solutions of a known dynamical system, the so called exosystem.

In this work, we consider a class of nonlinear output regulation problems where the global perspective plays an important role. More specifically, we study output regulation problems for rigid body systems, which are systems of particles with the holonomic constraint that the distance between any two particles is constant. Since this assumption holds for many mechanical systems, rigid body systems constitute one important class of mechanical models. One specific property of rigid bodies, which makes synthesis problems challenging, is the non-trivial geometry of the state space for the rotational motion of the rigid body. More precisely, continuous vector fields on the state space for the rotational motion with an asymptotically stable equilibrium will always have multiple equilibria. This is an obstruction for global stability of an equilibrium in a smooth vector field or for global stabilization of an equilibrium with smooth control laws. The synthesis of smooth and globally applicable control laws for this system class is important for many applications, for example robotic manipulators or multi-body systems. In this thesis, we present a novel observer based controller design methodology as a solution for the considered class of output regulation problems. Our approach is globally applicable and achieves convergence of every solution to one of the isolated components of the set of equilibria and that the desired equilibrium is the only asymptotically stable component while all other equilibria are unstable. The proposed approach is a two-step design method. The first step is the design of a state feedback, the second the design of an observer. We show that an output feedback controller consisting of the observer and the feedback which utilizes the estimated states solves the considered output regulation problem. An important property of our solution is that we analyze the convergence properties of the closed loop with the help of the convergence properties of the state feedback and the observer. More precisely, we show that the existence of a so called height function for the control system with the state feedback together with a convergent observer are sufficient to analyze the convergence properties in the output feedback case. Since this approach works for any convergent observer and any feedback provided that certain properties of the height function are fulfilled, our approach establishes a nonlinear separation principle for the considered problem class. This resembles the famous separation result for the output feedback problem in linear systems theory, where convergence is guaranteed despite the independent pole assignment for the state feedback and the observer. The presented two step design methodology and the separation result are not restricted to rigid body control problems and can be extended to similar setups.

Below, we give a brief outline of the thesis and summarize the main contributions of the thesis.

1.1 Outline and contributions

The thesis is structured with respect to the two considered topics, i.e. *synchronization in oscillator networks* in Chapter 2 and *global output regulation problems for rigid body systems* in Chapter 3. Both chapters contain two problems in connection with the considered topic. The chapters are structured as follows. In each chapter, we first give a general introduction to the considered problem class and review the necessary background literature. Then we consider each problem in a separate section. These sections start with a problem statement followed by additional background information. After this, we continue with the presentation of the results. We conclude each problem with a short summary. The chapters are then concluded by a summary and a discussion of the results. The conclusions of the thesis are presented in Chapter 4. In the following we give an outline of the chapters and briefly summarize the contributions of the respective chapters.

1.1.1 Chapter 2 — Synchronization of oscillators

Chapter 2 contains the material on synchronization in oscillator networks. The chapter is based on the publications Schmidt et al. [125; 126; 119]. We start Chapter 2 by a brief introduction to synchronization.

In Section 2.1, we give the theoretical background for our results on *synchronization in the Kuramoto model with delays* and for our results on our *Lyapunov based approach to synchronization in oscillator networks*.

Section 2.2 contains our results on synchronization in the Kuramoto model with delays. The main contributions are listed in the following.

- We give a result on state agreement for a network of single integrators with delayed coupling.

- We discuss a necessary conditions for frequency and phase synchronization in the Kuramoto model with delays.

- We give sufficient conditions for frequency synchronization in the Kuramoto model with delays with all-to-all coupling, which also yields a less conservative sufficient condition for non-delayed systems.

- We give a sufficient condition for phase and frequency synchronization for the Kuramoto model with delays where the graph which describes the coupling is directed and contains a directed spanning tree.

Section 2.3 contains our Lyapunov based approach to synchronization in oscillator networks. The main contributions are listed in the following.

- We characterize synchronization in oscillator networks by two new conditions.

- We give a sufficient condition for synchronization in a class of oscillator networks consisting of identical oscillators where we have a Lyapunov function for the periodic orbits of the uncoupled oscillators.

We conclude the chapter in Section 2.4.

1.1.2 Chapter 3 — Global output regulation and the regulation of rigid bodies

Chapter 3 contains the material on global output regulation for rigid body and rigid body like systems. The chapter is based on the publications Schmidt et al. [120; 122; 124], parts of Section 3.2 will appear in Schmidt et al. [123]. We start Chapter 3 with a brief introduction to the specific output regulation problem.

In Section 3.1 we give the general theoretical background on output regulation.

Section 3.2 contains our results for a class of output regulation problems on $SE(n)$. The main contributions are listed in the following.

- We present a two step procedure consisting of an observer design and a state feedback design. The certainty equivalence implementation of the feedback gives a solution for the considered class of output regulation problems.

- We show in the state feedback case that the closed loop solutions converge to the desired equilibrium for almost all initial conditions in the sense that the exceptional set of initial conditions has measure zero. Furthermore, we show in the output feedback case that every solution converges to one of the isolated components of the set of equilibria and the desired equilibrium is the only asymptotically stable component while all other equilibria are unstable. The convergence proof utilizes a height function associated with the state feedback.

- We establish a separation principle for the considered class of control systems. More precisely, the design of the state feedback and of the observer are independent and the convergence proof remains valid for any state feedback for which the properties of the height function are fulfilled and for any convergent observer.

Section 3.3 contains our results for a class of output regulation problems for rigid body systems. The main contributions are listed in the following.

- We adapt the two-step methodology presented in Section 3.2 for the class of rigid body control problems. This includes an extension of the feedback design and an extension of the observer design.

- We present an application scenario for the proposed design procedure for a multi-body satellite. Our presentation includes the detailed modeling and the detailed control design.

Section 3.4 contains convergence results for a differential equation on $SO(n)$ which were made publicly available in [121]. The main contributions are listed in the following.

- We show that the differential equation is a gradient flow on $SO(n)$ and give a detailed analysis of its convergence properties.

- We show that the set of initial conditions for which the solutions of a gradient flow of a Morse-Bott function converge to an unstable manifold of equilibria has measure zero.

We conclude the chapter in Section 3.5.

1.2 Mathematical background

In the following we summarize the mathematical background used without further explanations throughout the thesis. Additional background material is given wherever necessary.

1.2.1 Smooth maps and manifolds

We shortly summarize in the following what we mean by the terms *smooth function* and *smooth manifold* with the only goal to make the terms precise. The material is standard and, unless otherwise stated, the sources utilized in the following are Guillemin and Pollack [51], Milnor [97] and Lee [81].

We consider manifolds as subsets of an Euclidean space $\mathbb{R}^k$ such that the neighborhood of every point of this subset locally looks like $\mathbb{R}^n$. To make this precise we need the notion of smooth maps between subsets of Euclidean spaces.

Definition 1.1 (Guillemin and Pollack [51, Chapter 1, §1, p.2]). *A map $f : U \to \mathbb{R}^m$ for $U \subset \mathbb{R}^n$ and U open is called smooth if it has continuous partial derivatives of all orders.*

Definition 1.2 (Guillemin and Pollack [51, Chapter 1, §1, p.2]). *A map $f : \mathcal{M} \to \mathbb{R}^m$ for (an arbitrary) $\mathcal{M} \subset \mathbb{R}^n$ is called smooth if for every $x \in \mathcal{M}$ there is an open set $U \subset \mathbb{R}^n$ and a smooth map $F : U \to \mathbb{R}^m$ such that $F|_{U \cap \mathcal{M}} = f$.*

Definition 1.3 (Guillemin and Pollack [51, Chapter 1, §1, p.3]). *A smooth map $f : \mathcal{M} \to \mathcal{N}$ with $\mathcal{M} \subset \mathbb{R}^n$ and $\mathcal{N} \subset \mathbb{R}^m$ is a diffeomorphism if it is one-to-one and onto and the inverse map $f^{-1} : \mathcal{N} \to \mathcal{M}$ is smooth.*

Let $x \in U \subset \mathbb{R}^n$, $v \in \mathbb{R}^n$ and $\alpha : (-\varepsilon, \varepsilon) \to U$ be a differentiable curve with $\alpha(0) = x$ and $\frac{d}{dt}|_{t=0}\alpha(t) = v$. The derivative df_x of f at $x \in U$ in the direction v is given by

$$df_x(v) = \frac{d}{dt}|_{t=0}(f \circ \alpha)(t). \tag{1.1}$$

The map $df_x : \mathbb{R}^n \to \mathbb{R}^m$ defined by (1.1) is independent of the choice of curve, for details see e.g. Carmo [25, Chapter 2, Appendix].

Definition 1.4. *We call the map $df_x : \mathbb{R}^n \to \mathbb{R}^m$ defined by (1.1) the differential of f at x.*

We now define a manifold $\mathcal{M}$ in the same way as Guillemin and Pollack [51].

Definition 1.5 (Guillemin and Pollack [51, Chapter 1, §1, p.3]). *A smooth manifold $\mathcal{M}$ of dimension n is a subset of $\mathbb{R}^k$ with $k \geq n$ such that $\mathcal{M}$ is locally diffeomorphic to $\mathbb{R}^n$.*

The property that $\mathcal{M}$ is locally diffeomorphic to $\mathbb{R}^n$ means that for every $x \in \mathcal{M}$ there is a neighborhood $V \subset \mathcal{M}$ with $x \in V$ which is diffeomorphic to an open $U \subset \mathbb{R}^k$. A diffeomorphism $\phi : U \to V$ is called a *parametrization* of the neighborhood V. The inverse diffeomorphism $\phi^{-1} : V \to U$ is called a *coordinate system* on V. The pair (ϕ^{-1}, V) is called a *chart*. Let $x \in \mathcal{M}$, then we always find a parametrization ϕ such that $\phi(0) = x$. The *tangent space* $T_x\mathcal{M}$ of a manifold $\mathcal{M}$ at $x \in \mathcal{M}$ is the image of the map $d\phi_0 : \mathbb{R}^n \to \mathbb{R}^k$.

Definition 1.6 (Guillemin and Pollack [51, Chapter 1, §2, p.10]). *Let $\mathcal{M}$ and $\mathcal{N}$ be manifolds and let $f : \mathcal{M} \to \mathcal{N}$ be a smooth map. Furthermore, let $\phi : U \to \mathcal{M}$ be a parametrization of $\mathcal{M}$ around x and $\psi : V \to \mathcal{N}$ be a parametrization of $\mathcal{N}$ around $y = f(x)$ with $\phi(0) = x$ and $\psi(0) = y$ and let $h : U \to V$ be defined by $h = \psi^{-1} \circ f \circ \phi$. Then the differential df_x of $f : \mathcal{M} \to \mathcal{N}$ at $x \in \mathcal{M}$ is given by*

$$df_x = d\psi_0 \circ dh_0 \circ d\phi_0^{-1}. \tag{1.2}$$

For details on these terms and the independence of the differential and the tangent space of the parametrization, consult Guillemin and Pollack [51, Chapter 1].

The definition of a manifold as given in Definition 1.5 is not a principal restriction since every manifold $\mathcal{M}$ is diffeomorphic to a submanifold of a higher dimensional Euclidean space. More precisely, for every smooth manifold $\mathcal{M}$ of dimension n there is a smooth map $f : \mathcal{M} \to \mathbb{R}^{2n}$ such that f is injective, the differential $df_x : T_x\mathcal{M} \to T_{f(x)}\mathbb{R}^{2n}$ is injective for every $x \in \mathcal{M}$ and the inverse image $f^{-1}(K)$ of any compact $K \subset \mathbb{R}^{2n}$ is compact. This is the result of the Whitney embedding theorem, see e.g. Lee [81, Theorem 10.15], which implies that $f(\mathcal{M})$ is a smooth submanifold of $\mathbb{R}^{2n}$. In other words $\mathcal{M}$ is diffeomorphic to a smooth submanifold in $\mathbb{R}^{2n}$. A consequence of Definition 1.5 is that an open subset of $\mathbb{R}^n$ is a manifold and that a subspace of $\mathbb{R}^n$ is a manifold. In addition to that, the manifolds as given in Definition 1.5 are also smooth manifolds in the sense of Lee [81]. There, a smooth manifold is a *topological manifold* with a *smooth structure*. A *topological manifold* is a topological space with the Hausdorff property, which is second countable and which is locally homeomorphic to $\mathbb{R}^n$. A manifold from Definition 1.5 has the subspace topology, therefore $\mathcal{M}$ is Hausdorff (Lee [81, Lemma A.5(g)]) and second countable (Lee [81, Lemma A.5(h)]). Furthermore, a manifold from Definition 1.5 is locally diffeomorphic to $\mathbb{R}^n$ and since every differentiable map is also continuous, it is thus locally homeomorphic to $\mathbb{R}^n$. A *smooth structure* for a manifold $\mathcal{M}$ is a collection of charts $\{(\phi_i, U_i)\}$ such that (i) the union of the U_i covers $\mathcal{M}$, (ii) the charts are smoothly compatible, i.e. for any two charts (ϕ, U), (ψ, V) either $U \cap V = \emptyset$ or $\psi \circ \phi^{-1}$ is a diffeomorphism, and (iii) such that this collection of charts is maximal, i.e. it is not contained in any strictly larger collection of charts with the properties (i) and (ii). The property that $\mathcal{M}$ is locally diffeomorphic to $\mathbb{R}^n$ from Definition 1.5 directly implies the smooth compatibility of the charts.

In the context of this thesis, we need also the notions of *measure zero* and *dense* for subsets of a manifold $\mathcal{M}$. A set $A \subset \mathcal{M}$ of a manifold $\mathcal{M}$ is a set of measure zero if there is a collection of smooth charts $\{U_l, \phi_l\}$ whose domains cover A and such that $\phi_l(A \cap U_l)$ have measure zero in $\mathbb{R}^n$, i.e. the $\phi_l(A \cap U_l)$ can be covered for any ε by a countable collection of open balls whose volumes sum up to less than ε, for details see e.g. Lee [81, Chapter 10]. To define *dense*, we need the topological closure $\overline{A}$ of a set $A \subset \mathcal{M}$, i.e. the intersection of all closed sets in $\mathcal{M}$ that contain A. A dense subset of a smooth manifold $\mathcal{M}$ is a set $A \subset \mathcal{M}$ such that the topological closure fulfills $\overline{A} = \mathcal{M}$, see e.g. Lee [81, Appendix, Topology]. A is dense if and only if every nonempty open subset of X has non-empty intersection with A. The complement $\mathbb{R}^n \setminus A$ of a set of measure zero $A \subset \mathbb{R}^n$ is dense in $\mathbb{R}^n$, since if there is a point $x \in \mathbb{R}^n$ such that there is an open $U \subset \mathbb{R}^n$ with $x \in U$ and $U \cap (\mathbb{R}^n \setminus A) = \emptyset$, then A contains an open set and cannot have measure zero, see also Milnor [97, Chapter 2].

1.2.2 Differential equations

In the following we summarize the terminology and definitions from the theory of differential equations which are relevant for the whole thesis. The material is standard and, unless otherwise stated, the sources utilized in the following are Coddington and Levinson [31], Arnold [8], Hartman [56] and Chicone [28].

Autonomous differential equations

If not otherwise stated, $|x|$ for $x \in \mathbb{R}^n$ denotes the Euclidean norm of x. An autonomous ordinary differential equations is a differential equation of the form

$$\dot{x} = f(x), \tag{1.3}$$

where $f : M \to \mathbb{R}^n$ for a $M \subset \mathbb{R}^n$. We call M the *state space* and $x = (x_1, \ldots, x_n)$ the state of (1.3). We assume that f is smooth. A solution of the equation (1.3) with the initial condition $x_0 \in M$ is a function $x : I_{x_0} \to M, t \mapsto x(t; x_0)$ which fulfills the respective differential equation for all t in the interval $I_{x_0} = (t_\alpha(x_0), t_\omega(x_0)) \subset \mathbb{R} \neq \emptyset$ with $t_\alpha(x_0) < t_\omega(x_0)$ such that the solution fulfills $x(t_0; x_0) = x_0$. In the case (1.3) we can set without loss of generality $t_0 = 0$. Then $0 \in I_{x_0}$. We utilize the following definitions.

Definition 1.7 (Chicone [28, Section 1.8.1, Definition 1.66]). *A set $M' \subset M$ is positively invariant with respect to (1.3), if $x_0 \in M'$ implies that $x(t; x_0) \in M'$ for all $t \in I_{x_0}$.*

Definition 1.8 (Chicone [28, Section 1.1, p.2]). *The orbit of a solution $t \mapsto x(t; x_0)$ of the differential equation (1.3) with initial condition $x_0 \in M$ is the set*

$$\{x(t; x_0) \in M \mid t \in (t_\alpha(x_0), t_\omega(x_0))\}. \tag{1.4}$$

Definition 1.9 (LaSalle [80, Chapter 2, 2.2 Definition]). *Let $x_0 \in M$. A point $y \in M$ is called an ω-limit point of x_0 if there is a sequence $\{t_n\}$ with $t_n \in (t_\alpha(x_0), t_\omega(x_0))$ such that $t_n \to t_\omega$ for $n \to \infty$ and $x(t_n; x_0) \to y$ for $n \to \infty$. The ω-limit set $\omega(x_0)$ of x_0 is the union of its ω-limit points.*

Definition 1.10. *We say that a solution $t \mapsto x(t;x_0)$ remains bounded if $t_\omega(x_0) = \infty$ and if there is a constant $d \in \mathbb{R}$ such that $d > 0$ and $|x(t;x_0)| < d$ for all $t \in [0,\infty)$.*

An *equilibrium* $x_0 \in M$ of (1.3) is a point where $f(x_0) = 0$ holds. An important qualitative property of an equilibrium in the context of this thesis is stability, which always means stability in the sense of Lyapunov, defined in the following.

Definition 1.11 (Chicone [28, Section 1.6, Definition 1.38 and Definition 1.40]). *An equilibrium $x_0 \in M$ of (1.3) is called stable (or Lyapunov stable) if for every $\varepsilon > 0$ there is a $\delta(\varepsilon) > 0$ such that for every $y_0 \in M$ the condition $|x_0 - y_0| < \delta(\varepsilon)$ implies for all $t \geq 0$ the condition*

$$|x(t;x_0) - x(t;y_0)| < \varepsilon. \tag{1.5}$$

An equilibrium $x_0 \in M$ of (1.3) is called asymptotically stable and if there is an $a \in \mathbb{R}$, $a > 0$ such that $|y_0 - x_0| < a$ implies

$$\lim_{t \to \infty} |x(t;x_0) - x(t;y_0)| = 0. \tag{1.6}$$

For the analysis of qualitative properties of equilibria of (1.3) we utilize methods similar to the direct method of Lyapunov, see e.g. Chicone [28, Section 1.7]. Therefore we need the Lie derivative of a function (Lyapunov function) $V : M \to \mathbb{R}$.

Definition 1.12 (Arnold [8, Chapter 2, §10.2, Definition]). *Let $f : U \to \mathbb{R}^n$ with $U \subset \mathbb{R}^n$ be a vector field and $V : U \to \mathbb{R}$ a continuously differentiable function. The Lie-derivative $L_f V$ of V in the direction of the vector field f is defined by*

$$L_f V = \frac{\partial V}{\partial x} f. \tag{1.7}$$

In compliance with a large part of the literature we abbreviate $L_f V$ by $\dot{V}$.

Non-autonomous differential equations

A non-autonomous ordinary differential equations is a differential equation of the form

$$\dot{x} = f(x,t), \tag{1.8}$$

where $f : I \times M \to \mathbb{R}^n$ for $I \subset \mathbb{R}$ and $M \subset \mathbb{R}^n$. We assume that f is continuous in t and smooth in x. A solution of equation (1.8) with the initial condition $x_0 \in M$ is a function $x : I_{(t_0,x_0)} \to M, t \mapsto x(t;t_0,x_0)$ which fulfills the respective differential equation for all t in the interval $I_{(t_0,x_0)} = (t_\alpha(t_0,x_0), t_\omega(t_0,x_0)) \subset \mathbb{R} \neq \emptyset$ with $t_\alpha(t_0,x_0) < t_\omega(t_0,x_0)$ such that the solution fulfills $x(t_0;t_0,x_0) = x_0$. An equilibrium $x_0 \in M$ of (1.8) is a point where $f(t,x_0) = 0$ for all $t \geq t_0$.

Definition 1.13 (Iggidr and Sallet [62, Definition 2 and Definition 3]). *An equilibrium $x_0 \in M$ of (1.8) is called uniformly stable if for every $\varepsilon > 0$ there is a $\delta(\varepsilon) > 0$ such that for every $t_0 \in I$ and every $y_0 \in M$ the condition $|x_0 - y_0| < \delta(\varepsilon)$ implies that for all $t \geq t_0$*

$$|x(t;t_0,x_0) - x(t;t_0,y_0)| < \varepsilon. \tag{1.9}$$

An equilibrium is uniformly asymptotically stable if it is uniformly stable and there is an $a \in \mathbb{R}$, $a > 0$ such that for all $t_0 \in I$ and all y_0 such that $|x_0 - y_0| < a$ implies

$$\lim_{t \to \infty} |x(t;t_0,x_0) - x(t;t_0,y_0)| = 0 \tag{1.10}$$

uniformly with respect to t_0 and y_0.

We utilize a Theorem by Iggidr and Sallet [62], which we cite here for reasons of completeness.

Theorem 1.14 (Iggidr and Sallet [62, Theorem 5]). *Let $\dot{x} = f(x,t)$ be a system with a vector field which is locally Lipschitz in x uniformly in t. If on a neighborhood $\mathcal{O}_1$ of the origin there exists a continuously differentiable function $V : \mathcal{O}_1 \to \mathbb{R}$ such that*

- $V(x) \geq 0$ *for all $x \in \mathcal{O}_1$ and $V(0) = 0$.*

- $\dot{V}(x) \leq 0$ *for all $x \in \mathcal{O}_1$ and all $t \geq 0$.*

- *The restriction of f is uniformly asymptotically stable on the positively invariant set $\{x \in \mathcal{O}_1 | V(x) = 0\}$.*

Then the origin is an uniformly stable equilibrium point for $\dot{x} = f(x,t)$.

Convergence properties of differential equations

We utilize at several points a general result on the asymptotic behavior of solutions of ordinary differential equations from Arsie and Ebenbauer [9], which is repeated here for completeness. We consider a smooth manifold $\mathcal{M}$ with a given locally Lipschitz continuous vector field $\dot{x} = f(x)$. Consider an $x_0 \in \mathcal{M}$ such that a solution $t \mapsto x(t;x_0)$ is defined for all $t \in [0, \infty)$ and that the solution remains bounded. Assume that the omega-limit set $\omega(x_0) \in \mathcal{N} \subset \mathcal{M}$, where $\mathcal{N}$ is an embedded submanifold. Let $\mathcal{O}$ be an open neighborhood of $\mathcal{N}$ in $\mathcal{M}$.

Definition 1.15 (Arsie and Ebenbauer [9, Definition 1]). *A height function for the pair $(\mathcal{N}, f)$ is a continuously differentiable function $V : \mathcal{O} \to \mathbb{R}$ such that $\dot{V}|_{\mathcal{N}} \leq 0$ and $\dot{V}|_{\mathcal{N} \setminus \mathcal{E}} < 0$ where $\mathcal{E} = \{y \in \mathcal{N} | \dot{V} = 0\}$.*

Notice that $\dot{V}$ must only be negative semidefinite *on $\mathcal{N}$.* Let $\{\mathcal{E}_k\}$ be the connected components of $\mathcal{E}$.

Definition 1.16 (Arsie and Ebenbauer [9, Definition 5]). *Given a height function V, we say that the components $\{\mathcal{E}_k\}$ of $\mathcal{E} = \{y \in \mathcal{N} | \dot{V} = 0\}$ are contained in V if each $\mathcal{E}_k$ lies in a level set of V and the subset $\{V(\mathcal{E}_k)\} \subset \mathbb{R}$ has at most a finite number of accumulation points in $\mathbb{R}$.*

Theorem 1.17 (Arsie and Ebenbauer [9, Theorem 6]). *If the components $\{\mathcal{E}_k\}$ are contained in V, then $\omega(x_0) \subset \mathcal{E}_k$ for a unique k.* $\qquad\square$

2

Synchronization of oscillators

In the first part of this thesis we discuss problems which are connected to phenomena generally summarized under the term *synchronization*. The word synchronization originates from the Greek words σύν (*syn*, with), and χρόνος (*chronos*, time) and means *happening at the same time*, see e.g. Pikovsky et al. [111]. Loosely speaking, synchronization refers to a set of several processes where interactions between the processes results in an approximate or exact temporal alignment. A large variety of effects in the natural sciences and technology are labeled as synchronization phenomena. Some classical examples are listed in the following. The first well-known historical record is the pendulum synchronization of pendulum clocks mounted on a common beam as observed by Christiaan Huygens, see Huygens [61, p.243, No. 1335] or a translation of this letter in Pikovsky et al. [111, Appendix A.1]. John William Strutt (3rd Baron Rayleigh) described synchronization and oscillation death of spatially close organ pipes of the same pitch in his Theory of Sound, see Strutt [141] and Strutt [142] in §§ 322 c. The synchronization of triode generators by Appleton [5] was an early report of synchronization in electrical circuits. An often treated example in physics is the synchronization of lasers, see e.g. Roy and Thornburg [113]. Examples from biological sciences include the synchronization among cells of the sinoartrial pacemaker node, see e.g. Winfree [149, Chapter 14B], the synchronization of neurons in central pattern generators, see e.g. Ijspeert [63], and the synchronous flashing of fireflies described e.g. in Strogatz [140]. Fairly recent and broad surveys of synchronization are e.g. Pikovsky et al. [111] and Arenas et al. [6]. Synchronization problems from an engineering perspective are treated for example in Blekhman [18] and Nijmeijer and Rodriguez-Angeles [104]. During the last decade, synchronization was one of the phenomena studied within the research on so-called complex networks, see e.g. Strogatz [139] or Boccaletti et al. [19]. In this context, complex networks are dynamical systems which consist of a possibly large number of structurally similar, coupled dynamical systems. In systems and control, complex networks are considered under the term networked dynamic systems, see e.g. Olfati-Saber et al. [105].

In this thesis, we focus on synchronization in a specific class of networked dynamic systems. More precisely, we focus on synchronization in oscillator networks. An oscillator is a dynamical system with a periodic solution whose orbit has certain stability properties. An oscillator network is a large dynamical system consisting of coupled oscillators. In oscillator networks, synchronization means that the coupling between the oscillators drives the dynamics of the individual oscillators to the same oscillatory dynamics. In other words, synchronization in an oscillator network means that the dynamics of the network are similar to the dynamics of a single oscillator. The exact meaning depends on the oscillator models and the coupling mechanism. We consider here two different synchronization problems. The first problem deals with the effect of delays on synchronization in the Kuramoto model, which is a widely used model for synchronization. The second problem deals with more principal questions on the asymptotic dynamics in oscillator networks. The results of this chapter were published in Schmidt et al. [125; 119; 126].

This chapter is structured as explained in the following. In Section 2.1 we introduce the mathematical background for oscillators, oscillator networks and the Kuramoto model. Furthermore, we discuss the mathematical meaning of synchronization for these models. In Section 2.2 we present our results for the Kuramoto model with delays and discuss some implications of the results for undelayed models. In Section 2.3 we present our results on the asymptotic dynamics in oscillator networks. We conclude the chapter with a discussion of the results in Section 2.4. Additional literature for both results is reviewed in the specific sections.

2.1 Oscillators, oscillator networks and the Kuramoto model

In this section, we give the mathematical preliminaries on oscillators, oscillator networks and the Kuramoto model for the subsequent sections. This section is structured as explained in the following; we start by introducing oscillators as the basic building block of an oscillator network and subsequently define oscillator networks. After that, we explain a geometric picture associated with an oscillator network. We utilize this picture to explain the mathematics behind the synchronization problems considered in this chapter, namely the synchronization for delayed Kuramoto models and the Lyapunov-based approach for synchronization. The material on oscillators is covered in detail in books on ordinary differential equations, e.g. Chicone [28] and Teschl [143], or in special books on periodic motions e.g. Farkas [41]. Many details about the class of phase models, to which the delayed Kuramoto model belongs, are given e.g. in Hoppensteadt and Izhikevich [59].

2.1.1 Oscillators and oscillator networks

Subsequently, the distance $d : \mathbb{R}^n \times 2^{\mathbb{R}^n} \to \mathbb{R}$ between a set $A \subset \mathbb{R}^n$ and a point $x \in \mathbb{R}^n$ is the point-to-set distance with respect to the Euclidean metric, i.e.

$$d(x,A) = \inf_{y \in A} |x - y|. \tag{2.1}$$

A solution $t \mapsto x(t;x_0)$ of a differential equation $\dot{x} = f(x)$ with initial condition $x(0) = x_0$ is called a *periodic solution* if it is not an equilibrium and there is a $T > 0$ such that $x(t + T;x_0) = x(t;x_0)$ for all $t \geq 0$. The orbit $\{x(t;x_0)|t \in [0,T]\}$ (see Definition 1.8) associated with a periodic solution $t \mapsto x(t;x_0)$ is denoted by $\Gamma(x_0)$. Since for every $y_0 \in \Gamma(x_0)$, the orbit $\Gamma(y_0)$ is the same as $\Gamma(x_0)$, we drop the dependence on x_0 in $\Gamma(x_0)$. In the following, we define the stability concepts connected to periodic orbits. For details see e.g. Farkas [41, Chapter 5].

Definition 2.1 (Orbital stability and asymptotic orbital stability, Farkas [41, Definition 5.1.1 and Definition 5.1.2]). *Let $t \mapsto x(t;\gamma_0)$ be a periodic solution of $\dot{x} = f(x)$ with orbit Γ. A periodic orbit Γ is orbitally stable, if for every $\varepsilon > 0$ there is a $\delta(\varepsilon) > 0$ such that*

$$d(x_0,\Gamma) < \delta(\varepsilon) \Rightarrow \forall t \geq 0 : d(x(t;x_0),\Gamma) < \varepsilon. \tag{2.2}$$

A periodic orbit Γ is asymptotically orbitally stable if it is orbitally stable and there is a $\delta > 0$ such that

$$d(y_0, \Gamma) < \delta \Rightarrow \lim_{t \to \infty} d(x(t; y_0), \Gamma) = 0. \tag{2.3}$$

At some points we utilize a stricter stability property of the orbit:

Definition 2.2 (Exponential orbital stability). *Let $t \mapsto x(t; \gamma_0)$ be a periodic solution of $\dot{x} = f(x)$ with orbit Γ. A periodic orbit Γ is exponentially orbitally stable if there are constants $\delta, c_1, c_2 > 0$ such that*

$$d(x_0, \Gamma) < \delta \Rightarrow d(x(t; x_0), \Gamma) < d(x_0, \Gamma) c_1 e^{-c_2 t}. \tag{2.4}$$

The condition (2.4) immediately implies (2.2) and (2.3). Therefore, an exponentially stable periodic orbit is asymptotically stable. Note that the definition of orbital stability is the stability of the set Γ and does not include the time parametrization of the periodic solution. This means that orbital stability of a periodic orbit implies that a solution starting close to a periodic solution will only stay close to the associated periodic orbit. This does not mean that solutions starting close to a periodic orbit will stay close to the periodic solution. Furthermore, asymptotic orbital stability implies only that the ω-limit set of a solution starting close to Γ is given by Γ. Asymptotic stability of a periodic orbit Γ does not imply that a solution starting close to Γ converges to a solution $t \mapsto x(t; x_0)$ with $x_0 \in \Gamma$. Therefore, an additional property of a solution is introduced in the following.

Definition 2.3 (Asymptotic phase, Farkas [41, Definition 5.1.3]). *Let $t \mapsto x(t; \gamma_0)$ be a periodic solution of $\dot{x} = f(x)$ with orbit Γ. We say that $t \mapsto x(t; \gamma_0)$ has the asymptotic phase property if there is a $\delta > 0$ such that for every x_0 with $d(x_0, \Gamma) < \delta$ there exists a $\theta(x_0) \in \mathbb{R}$ such that*

$$\lim_{t \to \infty} |x(t + \theta(x_0); x_0) - x(t; \gamma_0)| = 0. \tag{2.5}$$

The asymptotic phase property for a periodic solution thus means, that every solution which converges to the orbit associated with the periodic solution also converges towards a periodic solution. To illustrate the asymptotic phase property we consider three examples.

Example 2.4 (Chicone [28, Exercise 2.134]). *Consider the system*

$$\begin{aligned} \dot{r} &= r(1 - r) \\ \dot{\phi} &= r, \end{aligned} \tag{2.6}$$

where the states r, ϕ fulfill $r \in \mathbb{R}_{>0}$ and $\phi \in [0, 2\pi)$, i.e. the system is given in polar coordinates. Separation of variables and integration shows that the solution $t \mapsto (r(t; (r_0, \phi_0)), \phi(t; (r_0, \phi_0)))$ of (2.6) for the initial (r_0, ϕ_0) is given by

$$\begin{aligned} r(t; (r_0, \phi_0)) &= \frac{r_0 \exp(t)}{1 - r_0 + r_0 \exp(t)} \\ \phi(t; (r_0, \phi_0)) &= \phi_0 + \log(1 - r_0 + r_0 \exp(t)) \quad \mathrm{mod}\ 2\pi. \end{aligned} \tag{2.7}$$

If $r_0 = 1$, the solution is periodic and the orbit is the unit circle. More precisely, $r_0 = 1$ implies $r(t; (1, \phi_0)) = 1$ and $\phi(t; (1, \phi_0)) = \phi_0 + t \mod 2\pi$ for any $\phi_0 \in [0, 2\pi)$ and all $t \geq 0$. Now let $\phi_0 \in [0, 2\pi)$ be arbitrary but fixed. Then any solution with $0 < r_0 \neq 1$ and $\tilde{\phi}_0 \in [0, 2\pi)$ fulfills

$$\lim_{t \to \infty} r(t; (r_0, \tilde{\phi}_0)) - r(t; (1, \phi_0)) = \lim_{t \to \infty} \frac{1}{1 + c_0 \exp(-t)} - 1 = 0. \tag{2.8}$$

13

Furthermore we have

$$\lim_{t \to \infty} \phi(t + \tau; (r_0, \tilde{\phi}_0)) - \phi(t; (1, \phi_0))$$

$$= \lim_{t \to \infty} \tilde{\phi}_0 + \log(1 - r_0 + r_0 \exp(t + \tau)) - \phi_0 - t \quad \mod 2\pi$$

$$= \tilde{\phi}_0 - \phi_0 + \lim_{t \to \infty} \log\left(\frac{1 - r_0}{\exp(t)} + r_0 \exp(\tau)\right) \quad \mod 2\pi. \tag{2.9}$$

If there is a τ such that $\log(r_0 \exp(\tau)) = \phi_0 - \tilde{\phi}_0$, then (2.9) implies that $\phi(t + \tau; (r_0, \tilde{\phi}_0)) - \phi(t; (1, \phi_0)) \to 0$ for $t \to \infty$, i.e. the asymptotic phase property of the unit circle which is the periodic orbit of (2.6). Since τ given by $\tau = \log(\frac{1}{r_0} \exp(\phi_0 - \tilde{\phi}_0))$ fulfills $\log(r_0 \exp(\tau)) = \phi_0 - \tilde{\phi}_0$, we see that the periodic orbit of (2.6) has the asymptotic phase property.

Note that the periodic orbit in Example 2.4 is exponentially orbitally stable, which is sufficient for the asymptotic phase property, see e.g. Chicone [28, Theorem 2.132]. However, exponential orbital stability of a periodic orbit is not necessary for the asymptotic phase property.

Example 2.5. *Consider the system*

$$\dot{r} = r(1 - r)^3$$
$$\dot{\phi} = 1, \tag{2.10}$$

where the states r, ϕ fulfill $r \in \mathbb{R}_{>0}$ and $\phi \in [0, 2\pi)$, i.e. the system is given in polar coordinates. The vector field for the r and the ϕ-part are independent, hence we can analyze them separately. We first analyze the equilibrium $r = 1$ for the r-part. Consider the function $V : \mathbb{R}_{>0} \to \mathbb{R}$ defined by $r \mapsto \frac{1}{2}(r - 1)^2$. Then we have

$$\dot{V}(r) = -r(1 - r)^4. \tag{2.11}$$

Since $r > 0$, this means that $\dot{V}$ is negative for all $r \neq 1$ and zero at $r = 1$. In other words the point $\bar{r} = 1$ is a globally asymptotically stable equilibrium for the r-dynamics. This means that the unit circle is a globally asymptotically stable periodic orbit for (2.10). As a consequence we have

$$\lim_{t \to \infty} r(t; (r_0, \tilde{\phi}_0)) - r(t; (1, \phi_0)) = 0. \tag{2.12}$$

The linearization of the r-dynamics around the equilibrium $\bar{r} = 1$ is given by

$$\frac{d}{dr}|_{r-1}(r(1 - r)^3) = ((1 - r)^3 + 3r(1 - r)^2)|_{r=1} - 0. \tag{2.13}$$

Hence the unit circle is not exponentially orbitally stable. Separation of variables and integration shows that the ϕ-component of the solution of (2.10) for the initial (r_0, ϕ_0) is given by

$$\phi(t; (r_0, \phi_0)) = t + \phi_0 \quad \mod 2\pi. \tag{2.14}$$

Since

$$\phi(t + \tau; (r_0, \tilde{\phi}_0)) - \phi(t; (1, \phi_0)) = t + \tau + \tilde{\phi}_0 - (t + \phi_0) \quad \mod 2\pi$$

$$= \tau + \tilde{\phi}_0 - \phi_0 \quad \mod 2\pi, \tag{2.15}$$

the choice $\tau = \phi_0 - \tilde{\phi}_0$ shows that we have the asymptotic phase property.

Example 2.6. *Consider the system*

$$\dot{r} = -(r-1)^3$$
$$\dot{\phi} = 1 + (r-1)^2, \tag{2.16}$$

where the states r, ϕ fulfill $r \in \mathbb{R}_{>0}$ and $\phi \in [0, 2\pi)$, i.e. the system is given in polar coordinates. Separation of variables and integration shows that the solution $t \mapsto (r(t;(r_0,\phi_0)), \phi(t;(r_0,\phi_0)))$ of (2.16) for the initial (r_0,ϕ_0) is given by

$$r(t;(r_0,\phi_0)) = 1 + \frac{r_0 - 1}{\sqrt{1 + 2t(r_0 - 1)^2}}$$
$$\phi(t;(r_0,\phi_0)) = t + \phi_0 + \frac{1}{2}\log(1 + 2(r_0 - 1)^2 t) \mod 2\pi. \tag{2.17}$$

If $r_0 = 1$, the solution is periodic and the orbit is the unit circle. More precisely, $r_0 = 1$ implies $r(t;(1,\phi_0)) \equiv 1$ and $\phi(t;(1,\phi_0)) = \phi_0 + t \mod 2\pi$ for any $\phi_0 \in [0,2\pi)$. Now let $\phi_0 \in [0,2\pi)$ be arbitrary but fixed. Then any solution with $0 < r_0 \neq 1$ and $\tilde{\phi}_0 \in [0,2\pi)$ fulfills

$$\lim_{t\to\infty} r(t;(r_0,\tilde{\phi}_0)) - r(t;(1,\phi_0)) = \lim_{t\to\infty} \frac{r_0 - 1}{\sqrt{1 + 2t(r_0 - 1)^2}} = 0. \tag{2.18}$$

Furthermore we have

$$\phi(t + \tau;(r_0,\tilde{\phi}_0)) - \phi(t;(1,\phi_0)) = \tilde{\phi}_0 + \tau + \frac{1}{2}\log(1 + 2(\tilde{r}_0 - 1)^2 t + \tau) - \phi_0 \mod 2\pi. \tag{2.19}$$

Since $\log$ is a monotonically increasing function, there is no τ such that the limit for $t \to \infty$ of equation (2.19) is zero. Therefore no point on the orbit $\Gamma = S^1$ has the asymptotic phase property. In other words, the limit set for any solution is the unit circle, but no solution which does not start on the unit circle converges to a periodic solution on the unit circle.

Remark 2.7. *In a comment after Definition 5.1.3 in Farkas [41], it is stated that the condition (2.5) is equivalent to*

$$\lim_{t\to\infty} |x(t;y_0) - x(t - \theta(y_0); \gamma_0)| = 0. \tag{2.20}$$

To see this, note that because the group property of a solution of a differential equation, see e.g. Chicone [28, Section 1.4], we have $x(t - \theta(y_0); \gamma_0) = x(t; x(-\theta(y_0); \gamma_0))$. If we set $\gamma(y_0) = x(-\theta(y_0); \gamma_0)$, then the asymptotic phase property implies that for every y_0 with $d(y_0, \Gamma) < \delta$ there is a $\gamma(y_0)$ such that

$$\lim_{t\to\infty} |x(t;y_0) - x(t; \gamma_0(y_0))| = 0. \tag{2.21}$$

On the other hand, assume that for every y_0 with $d(y_0, \Gamma) < \delta$ there is a $\gamma(y_0)$ such that (2.21) holds. Since $x_0, \gamma(y_0) \in \Gamma$, there is a $t(x_0)$ such that $x(t + t(x_0); x_0) = x(t; \gamma(y_0))$ for all t, which implies (2.20).

As a consequence, the asymptotic phase property is characterized by the statement that for every initial condition y_0 in a neighborhood of Γ, there is an initial condition $\gamma(y_0) \in \Gamma$ such that (2.21) holds. This means if a periodic solution with periodic orbit Γ has the asymptotic phase property, then every periodic solution with an initial condition on Γ has the asymptotic phase property. In other words the asymptotic phase property is a property of the periodic orbit associated with a periodic solution.

Utilizing this terminology, an oscillator and an oscillator network are defined in the following.

Definition 2.8 (Oscillator). *Consider a system of the form*

$$\dot{x} = f(x, u), \qquad (2.22)$$

where $f : \mathbb{R}^n \times \mathbb{R}^p \to \mathbb{R}^n$ is smooth, $x \in \mathbb{R}^n$ is the state and $u \in \mathbb{R}^p$ is the input. If the system (2.22) has for $u = 0$ an asymptotically orbitally stable periodic orbit Γ with the asymptotic phase property, then we call it an oscillator. An input affine oscillator is an oscillator of the form

$$\dot{x} = f(x) + g(x)u \qquad (2.23)$$

with a smooth $g : \mathbb{R}^n \to \mathbb{R}^{n \times p}$.

Definition 2.9 (Oscillator network, coupling functions). *Let $N \geq 2$ be a fixed natural number. An oscillator network is a system consisting of N (input affine) oscillators where $u_k = c_k(x_1, \ldots, x_N, \lambda)$ for $k \in \{1, \ldots, N\}$ and $c_k : \mathbb{R}^{nN} \times \mathbb{R}^l \to \mathbb{R}^p$ are smooth functions depending on the states $x_k \in \mathbb{R}^n$ and parameters $\lambda \in \mathbb{R}^l$. The c_k are called coupling functions.*

Remark 2.10 (Notation for the solutions of oscillator networks). *An oscillator network is a differential equations with state space $\mathbb{R}^{nN}$. We denote a solution for (2.34) with $x_0 \in \mathbb{R}^{nN}$ by $t \mapsto x(t; x_0) = (x_1(t; x_0), \ldots, x_N(t; x_0))^{\mathsf{T}}$. We call $t \mapsto x_k(t; x_0)$ the solution of oscillator k in the oscillator network.*

To obtain a geometric understanding for oscillator networks, we utilize that a periodic orbit Γ of a single oscillator is homeomorphic to the unit circle S^1. If Γ is the periodic orbit of an oscillator, then we always find coordinates such that in these coordinates $0 \notin \Gamma$, since Γ is non-trivial. In these coordinates, the map $\Gamma \to S^1$ defined by $x \mapsto \frac{x}{|x|}$ is a homeomorphism. As a consequence, the product of the limit cycles $\Gamma_1 \times \ldots \times \Gamma_N$ of the N uncoupled oscillators is homeomorphic to an N-torus $T^N = S^1 \times \ldots \times S^1$. Altogether, an oscillator network consists of a set of N oscillators and the product state space of these oscillators contains an invariant set $\mathcal{N} = \Gamma_1 \times \ldots \times \Gamma_N$ which looks topologically like the N-torus.

In synchronization problems for oscillators it is commonly assumed that the coupling preserves the existence of an invariant set $\mathcal{N}$ which is homeomorphic to the torus. Depending on the exact form of the coupling, this either means that this invariant set $\mathcal{N}$ is still $\Gamma_1 \times \ldots \times \Gamma_N$ or a slightly deformed version thereof. As loosely stated in the introduction, synchronization is achieved if the oscillators asymptotically have the same oscillatory dynamics. With the torus picture in mind, this means that a synchronous solution is a solution with a closed orbit on the invariant set $\mathcal{N}$. In the analysis of oscillator networks, the question is then whether this synchronous solution is stable or attractive. We consider here two different approaches to analyze the stability or attractivity of the synchronous solution, the phase model approach and the Lyapunov based approach. In the phase model approach, which is explained in the subsequent section (Section 2.1.2), one utilizes a model on the torus T^N to analyze the dynamics of the oscillator network in a neighborhood of the invariant set $\mathcal{N}$. The onset of synchronization in the oscillator network is then determined by the stability or attractivity of certain solutions of the phase model. In the Lyapunov based approach introduced in Section 2.1.3, we directly consider synchronization in oscillator networks as defined in Definition 2.9. Thereby we assume that the oscillators in the network are identical, that we have a Lyapunov function to check the asymptotic orbital stability of the periodic orbit Γ and that the coupling is zero if the difference of the states of any pair of oscillators is zero. This implies that

$\mathcal{N} = \Gamma_1 \times \ldots \times \Gamma_N = \Gamma \times \ldots \times \Gamma$ and that the synchronous solutions are the solutions consisting of N copies of a solution of the uncoupled oscillator. Since we utilize in our investigations the Lyapunov functions which show the orbital stability of the periodic orbit for the single oscillators, we call this in the following the *Lyapunov-based approach*.

2.1.2 Phase model approach and the Kuramoto model

In the phase model approach we analyze the dynamics of an oscillator network under the assumption that the coupling between the individual oscillators is small. Loosely speaking, this means that the model of the oscillator network is not significantly different from the model consisting of the N uncoupled oscillators. This implies that the invariant set $\mathcal{N}$ is not significantly different from $\Gamma_1 \times \ldots \times \Gamma_N$, i.e. $\mathcal{N}$ is still homeomorphic to a torus T^N. Furthermore, we can describe the dynamics of the oscillator network in a neighborhood of $\mathcal{N}$ by a simplified model on T^N, the so-called phase model. The main implication for synchronization is that it is possible to associate synchronization in the oscillator network with the attractivity of certain solutions in the phase model.

In the following we describe exactly what we mean by a phase model for the oscillator network. We start with a brief review of the phase model for a single system. Then we introduce the phase model for an oscillator network. After that we introduce two special solutions which play an important role for the investigation of synchronization in phase models.

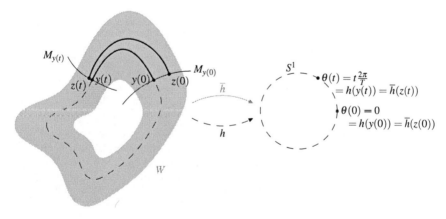

Figure 2.1: The dynamics of an oscillator on the periodic orbit Γ are described by the dynamics $\dot{\theta} = \omega$ on S^1, as illustrated in the Figure, which is adapted from Hoppensteadt and Izhikevich [59, Figure 4.5]. If the periodic orbit Γ is exponentially orbitally stable, then we know that there is a neighborhood W such that $W = \cup_{\gamma \in \Gamma} M_\gamma$ where $M_\gamma = \{y \in \mathbb{R}^n | \lim_{t \to \infty} |x(t; \gamma) - x(t; y)| = 0\}$. In other words, we can associate with every point $y \in W$ a point $\gamma(y) \in \Gamma$ such that the solution $t \mapsto x(t; y)$ through y converges asymptotically to the solution $t \mapsto x(t; \gamma(y))$. Thus the dynamics of an oscillator in a neighborhood of Γ are described by the dynamics of $\dot{\theta} = \omega$ on S^1.

The following arguments are along the lines of Hoppensteadt and Izhikevich [59, Section 4.1.7] and are illustrated by Figure 2.1, which is adapted from Hoppensteadt and Izhikevich [59, Figure 4.5]. Consider an *uncontrolled* (i.e. $u = 0$) oscillator of the form (2.22) with an exponentially orbitally stable periodic orbit Γ with period T. In the following, set $\omega = \frac{2\pi}{T}$. ω is called the *frequency* of the oscillator. Let $\gamma_0 \in \Gamma$ be fixed but arbitrary and denote the associated solution by $t \mapsto x(t; \gamma_0)$. For every point on $\gamma \in \Gamma$ there is exactly one $t \in [0, T)$ such that $\gamma = x(t; \gamma_0)$. Define $h : \Gamma \to S^1$ by $x(t; \gamma_0) \mapsto \frac{2\pi}{T} t = \omega t$. Since $t \mapsto x(t; \gamma_0)$ maps $[0, T)$ one-to-one on Γ, h is well defined and continuous. Therefore, h is a continuous map which maps periodic solutions of an uncontrolled ($u = 0$) oscillator (2.22) with orbit Γ onto solutions of

$$\dot{\theta} = \omega \qquad (2.24)$$

with $\theta \in S^1$. θ is called the *phase* of the oscillator. A specific property of exponentially orbitally stable oscillators is now, that the phase description is not only valid for solutions on Γ, but on a whole neighborhood W of Γ. To describe the dynamics of an uncontrolled ($u = 0$) oscillator (2.22) in a neighborhood W of Γ by (2.24), it is necessary to extend h to W. For an arbitrary $\gamma \in \Gamma$, define the set

$$M_\gamma = \{y \in \mathbb{R}^n | \lim_{t \to \infty} |x(t; \gamma) - x(t; y)| = 0\}. \qquad (2.25)$$

If Γ is exponentially orbitally stable, Guckenheimer [50] showed that there is a neighborhood W of Γ, such that $W = \cup_{\gamma \in \Gamma} M_\gamma$ and $M_{\gamma_1} \cap M_{\gamma_2} = \emptyset$ if $\gamma_1 \neq \gamma_2$. Let $\pi : W \to \Gamma$ denote the continuous function that maps each isochron to the associated point on the limit cycle, i.e. if $y \in W$ then $y \in M_\gamma$ for exactly one $\gamma \in \Gamma$ and then $\pi(y) = \gamma$. Guckenheimer [50] showed that $y_0 \in W \cap M_\gamma$ implies $\pi(x(t; y_0)) = x(t; \gamma)$ for all $t \geq 0$, in other words $x(t; y_0) \in M_{x(t; \gamma)}$ for all $t \geq 0$. This means that any solution starting in W converges to a solution on the periodic orbit, i.e. an exponentially stable orbit has the asymptotic phase property. The extension $\overline{h} : W \to S^1$ of h on W is defined by $h \circ \pi$. $\overline{h}$ maps solutions of (2.22) in W onto solutions of (2.24). An important consequence of the existence of $\overline{h}$ for this work is that we can analyze the dynamics of a single oscillator (2.22) in the vicinity of Γ by analyzing the model (2.24) on S^1.

In a similar fashion, it is possible to analyze the dynamics of some classes of oscillator networks by a model on the torus T^N. More specifically, consider an oscillator network of the form

$$\dot{x}_k = f_k(x_k) + \varepsilon c_k(x_1, \ldots, x_N, \varepsilon), \qquad (2.26)$$

where $k \in \{1, \ldots, N\}$ and the systems $\dot{x}_k = f_k(x_k)$ each have an exponentially stable periodic orbit Γ_k. This is a network of N input affine oscillators of the form (2.23) where $g_k(x_k) = \varepsilon I$ and where the coupling functions are smooth functions $c_k : \mathbb{R}^{nN} \times \mathbb{R} \to \mathbb{R}^n$ which depend on the parameter ε. We assume that the coupling is *weak*. In this context this means that $0 < \varepsilon \ll 1$ such that the vector field (2.26) is an ε-perturbation of the vector field

$$\dot{x}_k = f_k(x_k), \qquad (2.27)$$

where $k \in \{1, \ldots, N\}$. Under these assumptions, a theorem in Hoppensteadt and Izhikevich [59, Theorem 4.3] guarantees the existence of an invariant manifold $\mathcal{N}$ which is diffeomorphic to $\Gamma_1 \times \ldots \times \Gamma_N$. Furthermore, the theorem in Hoppensteadt and Izhikevich [59, Theorem 9.1] implies the existence of a neighborhood W of $\Gamma_1 \times \ldots \times \Gamma_N$ and a map $\overline{h} : W \to T^N$ which maps solutions of (2.26) to solutions of

$$\dot{\theta}_k = \omega_k + \varepsilon \tilde{c}_k(\theta_1, \ldots, \theta_N, \varepsilon), \qquad (2.28)$$

where $\theta_k \in S^1$ and the $\tilde{c}_k : T^N \times \mathbb{R} \to \mathbb{R}$ are smooth. ω_k is called the *frequency* of oscillator k. The theorem in Hoppensteadt and Izhikevich [59, Theorem 9.1] thus implies that it is possible to analyze the dynamics of an oscillator network of the form (2.26) in the vicinity of $\Gamma_1 \times \ldots \times \Gamma_N$ by analyzing (2.28). As explained at the end of Section 2.1.1, a synchronous solution is a solution with a closed orbit on T^N. In this thesis, we are interested in two specific types of synchronous solutions, their attractivity and the associated synchronization phenomenon. The first phenomenon ist *frequency synchronization*.

Definition 2.11 (Frequency synchronization in phase models)**.** *A phase model (2.28) achieves frequency synchronization for a $D \subseteq T^N$ if there is an $\Omega \in \mathbb{R}$ such that for every $\theta_0 \in D$ and every $k \in \{1, \ldots, N\}$ there are $\delta_1, \ldots, \delta_N$ with*

$$\lim_{t \to \infty} \theta_k(t; \theta_0) - \Omega t = \delta_k. \tag{2.29}$$

The frequency Ω is called agreement frequency.

Definition 2.11 implies that for every $k, i \in \{1, \ldots, N\}$ we have

$$\lim_{t \to \infty} \theta_k(t; \theta_0) - \theta_i(t; \theta_0) = \text{const}, \tag{2.30}$$

which is called *entrainment* by some authors, see e.g. [59, Section 9.1]. Frequency synchronization describes the effect that the coupling achieves that the oscillators with different frequencies ω_k asymptotically have the same frequency Ω. In Figure 2.2, we illustrate frequency synchronization for a phase model for two oscillators. The second phenomenon we

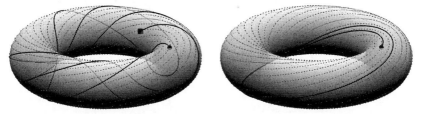

Figure 2.2: Illustrations for frequency synchronization in phase models. In both figures, we depict a part of the orbits of solutions of a two-dimensional phase model on the respective phase space, i.e. the torus $T^2 = S^1 \times S^1$. The black lines represent the depicted part of the orbits. The initial condition for each solution is indicated by a dot. The dotted lines are the orbits of frequency synchronized solutions, i.e. orbits of solutions where the phase of the first oscillator makes one rotation around S^1 while the phase of the second oscillator makes one rotation around S^1 and vice versa. The plot on the left side illustrates a part of an orbit in the case where the two oscillators are not coupled and where the natural frequencies of the oscillators are incommensurable. If the natural frequencies are incommensurable, then the orbit associated with the solutions is not closed and is a dense subset of the torus. The plot on the right side illustrates the orbits of two coupled oscillators. Since the coupling is strong enough, the solutions converge to a frequency synchronized solution which is associated with a closed orbit. The exact orbit depends on the initial condition.

consider is *phase synchronization*.

Definition 2.12 (Phase synchronization in phase models)**.** *A phase model (2.28) achieves phase synchronization if the system achieves frequency synchronization and in addition for all $k, i \in \{1, \ldots, N\}$ we have*

$$\lim_{t \to \infty} \theta_k(t; \theta_0) - \theta_i(t; \theta_0) = 0. \tag{2.31}$$

In Figure 2.3, we illustrate frequency synchronization for a phase model for two oscillators.

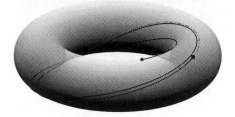

Figure 2.3: Illustration for phase synchronization in phase models. Similar to Figure 2.2, we depict a part of the orbit of a solution of a two-dimensional phase model on the respective phase space, i.e. the torus $T^2 = S^1 \times S^1$. The black line represents the depicted part of the orbit. The initial condition for the solution is indicated by a dot. The dash-dotted line is the orbit of the phase synchronized solution, i.e. the orbit of specific frequency synchronized solution. The coupling achieves that the two oscillators with the same natural frequency eventually converge towards the phase synchronized solution.

So far, we did not explain how to obtain the $\tilde{c}_k$ in (2.28) from the c_k in the oscillator network (2.26). We do not want to discuss this here, since it is a topic on its own, see for example [59, Chapter 9] or Ermentrout and Kopell [40]. Instead, we assume that the $\tilde{c}_k$ are given by

$$\tilde{c}_k(\theta_1,\ldots,\theta_N,\varepsilon) = \frac{K}{N\varepsilon} \sum_{l=1}^{N} a_{lk} \sin(\theta_l - \theta_k), \tag{2.32}$$

where $l \in \{1,\ldots,N\}$, $a_{lk} \in \{0,1\}$ and $K \in \mathbb{R}$. K is the so-called *coupling gain*. The phase model (2.28) with the coupling (2.32) given by

$$\dot{\theta}_k = \omega_k + \frac{K}{N} \sum_{l=1}^{N} a_{lk} \sin\left(\theta_l - \theta_k\right) \tag{2.33}$$

is the so called *Kuramoto model*. The Kuramoto model is one of the most investigated phase models to investigate synchronization. We give some additional background on the model in Section 2.2.1. In this work, in Section 2.2, we investigate the impact of delays in the coupling functions of the form (2.32) on frequency and phase synchronization.

2.1.3 Lyapunov-based approach

In contrast to the phase model approach, the Lyapunov-based approach is a non-local approach to synchronization in oscillator networks. This means that the analysis is not restricted to a neighborhood of $\Gamma_1 \times \ldots \times \Gamma_N$. The starting point of our investigation is again the informal description that synchronization in an oscillator network is achieved if all oscillators converge to the same periodic solution which is a solution of an individual oscillator.

For the Lyapunov based approach, we consider oscillator networks consisting of N identical oscillators. This means that the uncoupled oscillators all have the same periodic orbit Γ. In this setting, we define synchronization as follows:

Definition 2.13 (Synchronization of identical oscillators). *Let $N \in \mathbb{N}$ and $N \geq 2$. Consider an oscillator network of the form*

$$\dot{x}_k = f(x_k, c_k(x_1, \ldots, x_N)). \tag{2.34}$$

Furthermore, let Γ denote a periodic orbit of $\dot{x}_k = f(x_k, 0)$ for all $k \in \{1, \ldots, N\}$ and let $D \subset \mathbb{R}^{nN}$. We say that the oscillators synchronize in $D \subset \mathbb{R}^{nN}$ if there exists a $\gamma \in \Gamma$ such that for all $k \in \{1, \ldots, N\}$ and all $x_0 \in D$ we have $x_k(t; x_0) \to x_k(t; \gamma)$ for $t \to \infty$.

Synchronization in the sense of Definition 2.13 means that the solutions starting in D converge to a solution which consists of N copies of a solution of the uncoupled oscillator. As explained in Section 2.1.2, we can always define the notion of a phase for a single oscillator on the periodic orbit Γ. Then Definition 2.13 means that the synchronous solutions are the ones where all oscillators asymptotically reach the same phase. In other words, synchronization in the sense of Definition 2.13 implies phase synchronization (see Definition 2.12). One difference to the phase model approach is that we do not require exponential orbital stability of the periodic orbits of the single oscillators, which is only sufficient but not necessary for the asymptotic phase property of the periodic orbits, see also Example 2.5.

Note that the condition in Definition 2.13 implies for all $k, i \in \{1, \ldots, N\}$ and all $x_0 \in D$ that we have $x_k(t; x_0) - x_i(t; x_0) \to 0$ for $t \to \infty$. Consider the set $\mathscr{S}$ defined by

$$\mathscr{S} = \{(x_1, \ldots, x_N)^{\mathsf{T}} \in \mathbb{R}^{nN} | x_i = x_j \text{ for all } i, j \in \{1, \ldots, N\}\}. \tag{2.35}$$

Then the condition that for all $k, i \in \{1, \ldots, N\}$ and all $x_0 \in D$ we have $x_k(t; x_0) - x_i(t; x_0) \to 0$ for $t \to \infty$ is equivalent to the property that every solution converges towards $\mathscr{S}$. If we choose an initial condition for the oscillator network on $\mathscr{S}$, where the coupling function is zero, then the dynamics are described by the set of N uncoupled oscillators. In other words, the oscillator network dynamics are given by $\dot{x}_k = f(x_k, 0)$ for $k \in \{1, \ldots, N\}$. If we assume that the periodic orbits Γ are the only asymptotically stable limit sets for $\dot{x}_k = f(x_k, 0)$, this would imply that all oscillators converge to the same periodic solution. The orbit of this solution is then the intersection of $\Gamma \times \ldots \times \Gamma$ with $\mathscr{S}$. Therefore, solutions that start in $\mathscr{S}$ trivially achieve synchronization as defined in 2.13. The question is now whether convergence to $\mathscr{S}$ implies synchronization in the sense of Definition 2.13. In Section 2.3, we discuss why convergence to $\mathscr{S}$ does in general not imply synchronization. Furthermore, we utilize the assumption that there is a Lyapunov function for the periodic orbits of the individual oscillators to give sufficient conditions for synchronization.

2.2 Synchronization in the Kuramoto model with delays

In this section we consider synchronization problems for the Kuramoto model with delays. We investigate the influence of time delays on synchronization. We study the two synchronization phenomena *frequency synchronization* and *phase synchronization*.

We provide a lower bound on the coupling gain such that frequency synchronization is achieved in a network of non-identical Kuramoto oscillators in the case of all-to-all coupling. Our results include constraints on the initial conditions of the states of the system which result from the use of Lyapunov Razumikhin arguments for nonlinear retarded functional differential equations. In addition, we show that frequency synchronization and phase synchronization is possible in the case of non-identical Kuramoto oscillators with delays. Phase synchronization cannot be achieved in the case of non-identical natural frequencies without delays, for details see e.g. Section 2.2.3. The reason for this is a self-consistency condition which relates the agreement frequency, the natural frequencies, and the delays. The contributions presented in this Section were published in Schmidt et al. [125] and Schmidt et al. [126].

The remainder of this Section is structured as follow; we start with a problem statement in 2.2.1. Subsequently, we give a brief literature review on the Kuramoto model and the delayed Kuramoto model in Section 2.2.2. Furthermore, we discuss some preliminaries in Section 2.2.2. In Section 2.2.3, we present necessary conditions for frequency synchronization and phase synchronization and in Section 2.2.4, we present sufficient conditions for frequency synchronization and phase synchronization. We collect the technical proofs of Section 2.2 in Section 2.2.5 and give a summary in Section 2.2.6.

2.2.1 Problem statement

The subject of study in the present section is a network of delayed Kuramoto oscillators given by N delay differential equations of the form

$$\dot{\theta}_i(t) = \omega_i + \frac{K}{d_i} \sum_{k=1}^{N} a_{ki} \sin\left(\theta_k(t - \tau_{ki}) - \theta_i(t)\right), \tag{2.36}$$

where $i \in \{1, \ldots, N\}$, $\theta_i \in S^1$ denotes the phase, ω_i the natural frequency of system i and K the coupling gain. The $a_{ki} \in \{0, 1\}$ describe the coupling, i.e. $a_{ki} = 1$ if oscillator k is connected to system i and $a_{ki} = 0$ otherwise. The $d_i = \sum_k a_{ki}$ denote the number of systems connected to system i. Finally, $0 \leq \tau_{ki} \in \mathbb{R}$ are the constant delays between the systems k and i. We assume that $K > 0$ and $\tau_{ii} = 0$.

We study frequency synchronization and phase synchronization for the delayed Kuramoto model. In contrast to the solutions of ordinary differential equations, we denote solutions of (2.36) by $t \mapsto (\theta_1(t; \phi), \ldots, \theta_N(t; \phi))^{\mathsf{T}}$ where the initial ϕ is a function since the right hand side of (2.36) does not only depend on the value of the state $\theta = (\theta_1, \ldots, \theta_N)^{\mathsf{T}}$ at time t but also on the values of θ at $t - \tau_{ki}$ for all $k, i \in \{1, \ldots, N\}$. We make this more precise in Section 2.2.2 (Background for functional differential equations). For the problem statement it is sufficient to consider the solutions as functions of time. Then frequency synchronization in the sense of Definition 2.11 means there is a set of initial conditions Ξ and an Ω such that for every $\phi \in \Xi$ and every $k \in \{1, \ldots, N\}$ there are $\delta_1, \ldots, \delta_N$ with

$$\lim_{t \to \infty} \theta_k(t; \phi) - \Omega t = \delta_k. \tag{2.37}$$

Similarly, phase synchronization in the sense of Definition 2.12 for the delayed Kuramoto model means frequency synchronization and for all $k, i \in \{1, \ldots, N\}$

$$\lim_{t \to \infty} \theta_k(t; \phi) - \theta_i(t; \phi) = 0. \tag{2.38}$$

The goal is to derive conditions for these two synchronization effects in dependence on the system parameters.

Problem statement 2.14. *Consider the delayed Kuramoto model* (2.36).

    a) Assume that for all $k, i \in \{1, \ldots, N\}$ we have $a_{ki} = 1$. Find sufficient conditions for frequency synchronization in the model (2.36).

b) *Find sufficient conditions for phase and frequency synchronization without the restriction from a) in the model* (2.36).

Before presenting our results, we give a short literature overview and some mathematical preliminaries in the following section.

2.2.2 Preliminaries

In this section, we first give a brief review of the relevant literature and then a brief account of the mathematical background for our results on the delayed Kuramoto oscillators. The delayed Kuramoto oscillators belong to the class of *functional differential equations*. Since some properties are different for this class of differential equations, we summarize these properties shortly in the following section. In addition to that, we review very shortly the necessary terminology from graph theory, which we utilize in our results.

The Kuramoto model without delays (2.33) is one of the most studied models in connection with oscillator synchronization. For a general overview on the research concerning the Kuramoto oscillator and its importance for applications, see Strogatz [138], Acebrón et al. [1], Dörfler and Bullo [35]. In recent years a fair amount of contributions showed that the use of control theoretic methods helps to understand the synchronization effects and find new application scenarios for the Kuramoto model, see e.g. Aeyels and Rogge [2], Jadbabaie et al. [70], Lin et al. [85], Verwoerd and Mason [146], Chopra and Spong [30], Sarlette and Sepulchre [117]. Two particularly interesting application areas for the Kuramoto model from the control perspective are power networks see e.g. Dörfler et al. [36], and vehicle coordination problems, see e.g. Sepulchre et al. [129] and Sepulchre et al. [130].

Kuramoto oscillators with delays were considered in numerous publications starting from two oscillators with symmetric delays, e.g. Schuster and Wagner [127], to oscillator networks with homogeneous delays, see e.g. Nakamura et al. [103], Luzyanina [87], Yeung and Strogatz [151], Choi et al. [29], Earl and Strogatz [38], Münz et al. [101]. A mathematical justification of explicit time-delays in phase models in general and the Kuramoto model in particular was given some time after the first contributions, see Izhikevich [69]. This contribution shows that explicit time-delays appear in phase models if the delays are much longer than the period of the individual oscillators. More recently, Kuramoto oscillators with constant and heterogeneous delays were studied in the continuum limit, i.e. for an infinite number of oscillators, for details see Lee et al. [82]. Furthermore, a linear stability analysis of a network with constant and heterogeneous delays was performed in Papachristodoulou and Jadbabaie [106]. In contrast to most of the physics literature, we consider arbitrary large but finite populations of oscillators. Furthermore, we consider non-identical oscillators with constant and heterogeneous delays and give an analysis of frequency synchronization for all-to-all coupling and phase synchronization for directed networks with a spanning tree. The possibility of phase synchronization in Kuramoto oscillators with nonidentical natural frequencies and undirected communication graphs was already observed in Papachristodoulou et al. [107]. However, the focus of Papachristodoulou et al. [107] was the effect of delays on the functionality of large-scale multi-agent systems with nonlinear dynamics, where the persistence of functionality could be guaranteed even in presence of switching topologies. In contrast, we focus on frequency synchronization and phase synchronization for delayed Kuramoto oscillators and develop a unified framework for both problems.

Background for functional differential equations

Since the Kuramoto model with delays given by (2.36) is not an ordinary differential equation, but a delay differential equation, we give some additional mathematical background information. We utilize this notation only throughout Section 2.2. The following definitions are taken from Hale and Lunel [55] and Haddock and Terjeki [52] and adapted for our purpose.

As already indicated in Section 1.2, let $|\cdot|$ denote the Euclidean norm of $x \in \mathbb{R}^n$. For $[a,b] \subset \mathbb{R}$ let $C([a,b],\mathbb{R}^n)$ denote the space of continuous functions from $[a,b]$ to $\mathbb{R}^n$. For $0 \le T \in \mathbb{R}$, define C_T as the set $C([-T,0],\mathbb{R}^n)$, where C_T is equipped with the norm $||\phi||_\infty = \sup_{-T \le s \le 0} |\phi(s)|$ for $\phi \in C_T$. For $x \in C([-T,\infty),\mathbb{R}^n)$ and any $t \ge 0$ we define $x_t(s) = x(t+s)$ for all $-T \le t \le 0$, i.e. x_t is the restriction of x to $[-T+t,t]$.

Let Ξ be a subset of C_T, $f : \Xi \to \mathbb{R}^n$ a given function, and let "$\dot{}$" represent the right-hand derivative, then we call

$$\dot{x}(t) = f(x_t) \tag{2.39}$$

an autonomous *Retarded Functional Differential Equation* (RFDE) on Ξ. A simple class of RFDEs are the so called *differential difference equations* which are of the form

$$\dot{x}(t) = f(x(t), x(t-T)) \tag{2.40}$$

for a $T > 0$. For a given $\phi \in \Xi$, we say $x : [-T,\rho) \to \mathbb{R}^n$ is a *solution of* (2.39) on $[-T,\rho)$ with initial condition $\phi \in C_T$ for a $\rho > 0$, if $x \in C([-T,\rho),\mathbb{R}^n)$, $x_t \in \Xi$ and x satisfies (2.39) for all $t \in [0,\rho)$. For a solution with initial condition ϕ we write $t \mapsto x(t;\phi)$. Then, $s \mapsto x_t(s;\phi)$ is the restriction of $s \mapsto x(s;\phi)$ to $[-T+t,t]$. Without loss of generality we assume that $f(0) = 0$. Then $t \mapsto x(t;0)$ with $x(t;0) = 0$ for all $t \ge 0$ is a solution of (2.39).

We adapt some of the definitions of invariance and stability from the ordinary differential equation context for RFDEs. The following definitions are all taken from Haddock and Terjeki [52].

Definition 2.15. *We call $M \subset C_T$ an invariant set with respect to an RFDE (2.39), if for any $\phi \in M$, there is a solution $t \mapsto x(t;\phi)$ defined on $(-\infty,\infty)$ such that $s \mapsto x_t(s;\phi) \in M$ for all $t \in (-\infty,\infty)$. A set $M \subseteq \Xi$ is said to be positively invariant if, for any $\phi \in M$, $s \mapsto x_t(s;\phi)$ for all $t \ge 0$.*

Definition 2.16. *We say that $\psi \in C_T$ is in the ω-limit set $\omega(\phi)$ of a solution $t \mapsto x(t;\phi)$ of (2.39) if there is a sequence $\{t_n\}$ with $t_n \ge 0$, $t_n \to \infty$ and $||x_{t_n}(s;\phi) - \psi||_\infty \to 0$ for $n \to \infty$.*

Assume $t \mapsto x(t;\phi)$ is a solution of (2.39) that is defined and bounded on $[-T,\infty)$, then the set $\{x_t(s;\phi) | t \ge 0\}$ is precompact, $\omega(\phi)$ is non-empty, compact, connected, and invariant, and $x_t(s;\phi) \to \omega(\phi)$ as $t \to \infty$, see Hale and Lunel [55, Chapter 3].

Definition 2.17. *The solution $t \mapsto x(t;0)$ with $x(t;0) = 0$ for all $t \ge 0$ is stable, if for every $s \in \mathbb{R}$, $\varepsilon > 0$ there is a $\delta(\varepsilon) > 0$ such that $||\phi||_\infty < \delta$ implies $||x_t(s;\phi)||_\infty < \varepsilon$ for $t \ge 0$. The solution $t \mapsto x(t;0)$ for $t \ge 0$ is asymptotically stable if it is stable and there is a $\delta > 0$ such that $||\phi||_\infty < \delta$ implies that $x(t;\phi) \to 0$ for $t \to \infty$.*

Graph Theory

In the following, we consider models of the form (2.33) and (2.36). These are models of coupled differential equations in the case of (2.33) or coupled delay differential equations in the case of (2.36), respectively. It is possible to describe the coupling structure in these models with a graph. In recent years this description of the interaction structure in coupled systems was established especially in the context of networked dynamic systems, see e.g. Olfati-Saber et al. [105]. One of the main advantages of this description is that it is possible to associate a matrix with the graph, whose spectral properties imply properties about the networked dynamical system and about the structure of the graph, see also Olfati-Saber et al. [105]. In this thesis, we do not take advantage of the spectral properties of a matrix which is associated to the graph, but we utilize the language from graph theory to describe the interaction structure in models of the form (2.33) and (2.36). This makes it easier to present the given statements. We adapt here a standard terminology and refer to Godsil and Royle [48] for more details.

Definition 2.18. *Let $N \in \mathbb{N}$. A directed graph $G = (V(G), E(G))$ consists of a set $V(G)$ of N elements which are called vertices (nodes) and a set $E(G) \subseteq V(G) \times V(G)$ whose elements are called edges.*

Note that we utilize $V(G)$ to make the distinction to Lyapunov functions which are in general also denoted by V.

Definition 2.19. *The adjacency matrix $A(G) \in \mathbb{R}^{N \times N}$ of G is defined by $a_{ij} = 1$ if $e_{ij} \in E(G)$ and $a_{ij} = 0$ if $e_{ij} \notin E(G)$.*

Therefore a matrix $A \in \{0,1\}^{N \times N}$ characterizes a directed graph. If we consider the $a_{ki} \in \{0,1\}$ in (2.33) or (2.36) as the entries of an adjacency matrix, then we have a graph $G(A)$ which is associated with (2.33) or (2.36), respectively. For each oscillator there is a phase which is associated to a single node. If $e_{ij} = (v_i, v_j) \in E(G)$, then there is an edge from node v_i to node v_j. If there is an edge from node v_i to node v_j, oscillator i is coupled to oscillator j. We summarize in the following definition some terminology that we utilize later.

Definition 2.20. *Let $G = (V(G), E(G))$ be a directed graph.*

a) If $e_{ij} = (v_i, v_j) \subset E(G)$, then we call v_i a parent of v_j.

b) A directed path from v_i to v_j is a sequence of edges out of $E(G)$ of the form (v_{i_0}, v_{i_1}), $(v_{i_1}, v_{i_2}), \ldots, (v_{i_{p-1}}, v_{i_p})$ where $\{i_k\}_{0 \le k \le p}$ is a finite sequence with $i_k \in \{1, \ldots, N\}$ and $i_0 = i$ and $i_p = j$.

c) A directed tree is a directed graph where there is a node, the so called root, such that there is exactly one directed path from the root to every vertex of the graph.

d) A subgraph $(V(G)', E(G)')$ of G is a directed graph with $V(G)' \subseteq V(G)$ and $E(G)' \subseteq E(G)$.

e) If there exists a subgraph $(V'(G), E(G)')$ of G that is a directed tree, then we say that G contains a directed spanning tree. We denote the set of all roots of G as $V(G)_R$.

In the following, we also say spanning tree when referring to a directed spanning tree.

We frequently utilize one special graph in the following, which is defined in the following.

Definition 2.21. *We say that the graph is an all-to-all graph if $a_{ki} = 1$ for all $k, i \in \{1, \ldots, N\}$.*

2.2.3 Necessary conditions for synchronization

In the following we consider necessary conditions for frequency and phase synchronization. We show that the agreement frequency is characterized by one equation, the so called self-consistency condition. Its implications for the undelayed case were a focus of several publications Kuramoto and Nishikawa [76], Strogatz [138], Aeyels and Rogge [2], Franci et al. [43]. The self-consistency condition for phase synchronization in the delayed case was already used in Papachristodoulou et al. [107]. There, it was part of the sufficient condition to achieve phase synchronization. In this section, we will discuss the details of the self-consistency condition in the delayed case for both, frequency synchronization and phase synchronization.

From Definition 2.11, synchronization implies that all oscillators asymptotically agree on a agreement frequency Ω. Therefore we obtain the following condition that for all $i \in \{1,\ldots,N\}$ there is a $\delta_i \in \mathbb{R}$ such that

$$\lim_{t \to \infty} \theta_i(t;\phi) - \Omega t = \delta_i, \tag{2.41}$$

where δ_i is an offset depending on the communication delays τ_{ki}, the natural frequencies ω_i, the coupling constant K and the graph topology. Substituting the limit of expression (2.41) into (2.36), gives the equation

$$\Omega = \omega_i + \frac{K}{d_i} \sum_{k=1}^{N} a_{ki} \sin\left(\Omega(t - \tau_{ki}) + \delta_k - \Omega t - \delta_i\right)$$

$$= \omega_i + \frac{K}{d_i} \sum_{k=1}^{N} a_{ki} \sin\left(-\Omega \tau_{ki} + \delta_k - \delta_i\right). \tag{2.42}$$

Equation (2.42) is called *self-consistency condition for frequency synchronization*, since it characterizes the agreement frequency the oscillators have to agree on, if they synchronize. Since neither the δ_i nor Ω are known from the parameters of the system (2.36), it is not straight forward to decide in which case (2.42) has a solution. In Theorem 2.23 in Section 2.2.4, we show that frequency synchronization is possible in delayed Kuramoto models if the associated graph is an all-to-all graph without assuming that (2.42) has a solution. Therefore we implicitly give conditions for the all-to-all case which assure that (2.42) has a solution.

Phase synchronization means that the phase difference vanishes in the limit. Therefore the oscillators have to evolve with an agreement frequency, i.e. we have the condition that there is a $\delta \in \mathbb{R}$ such that for all $i \in \{1,\ldots,N\}$ we have

$$\lim_{t \to \infty} \theta_i(t,\phi) - \Omega t = \delta. \tag{2.43}$$

Substituting expression (2.43) into (2.36) gives the equations

$$\Omega = \omega_i - \frac{K}{d_i} \sum_{k=1}^{N} a_{ki} \sin(\Omega \tau_{ki}), \tag{2.44}$$

for all $i \in \{1,\ldots,N\}$. We call equation (2.44) the *self-consistency condition for phase synchronization*. The importance of (2.44) becomes clear if one considers the undelayed case. If $\tau_{ki} = 0$ we obtain from (2.44) the following equations

$$\Omega = \omega_i - \frac{K}{d_i} \sum_{k=1}^{N} a_{ki} \sin(\Omega \tau_{ki}) \overset{\tau_{ki}=0}{=} \omega_i, \tag{2.45}$$

for all $i \in \{1, \ldots, N\}$. This means that the oscillators need to have the same natural frequencies for phase synchronization to occur. In other words, in the undelayed case we can observe phase synchronization only if we have identical oscillators. In the delayed case the agreement frequency is characterized by equation (2.44). We show in Section 2.2.3 that the existence of a solution Ω of (2.44) forms a part of the sufficient conditions for phase synchronization. For a given graph associated to the model it is therefore possible to check the existence of a phase synchronized solution. However, searching for suitable parameters such that (2.44) has a solution Ω is in general difficult. For example, for a given oscillator network with known natural frequencies and known topology, it is in general not clear how to choose the delays to guarantee the existence of a phase synchronized solution.

In case a solution of (2.44) is known, it is possible to investigate the dependence on the model parameters by utilizing the *implicit function theorem* which gives sufficient conditions when it is possible to locally invert a function with respect to some of its variables, see e.g. Rudin [114]. More specifically, we have the graph $G = (V(G), E(G))$ associated with the Kuramoto model and consider the function $f : \mathbb{R}^N \times \mathbb{R} \times \mathbb{R}^{|E(G)|} \to \mathbb{R}^N$ for all $i, k \in \{1, \ldots, N\}$ defined by

$$f(\omega_1, \ldots, \omega_N, \Omega, \tau_{1,1}, \ldots, \tau_{1,N}, \ldots, \tau_{N,N}) = \begin{bmatrix} \Omega - \omega_1 + \frac{K}{d_1} \sum_{k=1}^{N} a_{k1} \sin(\Omega \tau_{k1}) \\ \vdots \\ \Omega - \omega_N + \frac{K}{d_N} \sum_{k=1}^{N} a_{kN} \sin(\Omega \tau_{kN}) \end{bmatrix}. \qquad (2.46)$$

f is differentiable. A set of parameters $\overline{\omega}_i, \overline{\Omega}, \overline{\tau}_{ki}$, where $i, k \in \{1, \ldots, N\}$ and which solve the equation (2.44), fulfill $f(\overline{\omega}_1, \ldots, \overline{\omega}_N, \overline{\Omega}, \overline{\tau}_{1,1}, \ldots, \overline{\tau}_{1,N}, \ldots, \overline{\tau}_{N,N}) = 0$. We want to determine quantitatively in which way we are able to vary N parameters in dependence of the remaining $N + |E(G)| + 1 - N = |E(G)| + 1$ parameters such that we still have a solution of (2.44). Therefore, let $(\overline{x}, \overline{y})$ be a partition of the set of parameters $\overline{\omega}_i, \overline{\Omega}, \overline{\tau}_{ki}$ such that $\overline{x} \in \mathbb{R}^{|E(G)|+1}$ and $\overline{y} \in \mathbb{R}^N$. We consider $f(\omega_1, \ldots, \omega_N, \Omega, \tau_{1,1}, \ldots, \tau_{1,N}, \ldots, \tau_{N,N})$ as $\overline{f}(x, y)$, then $f(\overline{x}, \overline{y}) = 0$. If the Jacobian of $\overline{f}$ with respect to y at $\overline{y}$ is regular, then the implicit function theorem guarantees the existence of a neighborhood U of $\overline{x}$, a neighborhood V of $\overline{y}$ and a locally defined function $g : U \to V$ such that $\overline{f}(x, g(x)) = 0$ for all $x \in U$. More specific consideration of the parameter dependence of existing solutions to (2.44) require an additional research effort by either considering specific graph topologies or by utilization of a theory which can handle non-differentiable mappings to cope with changes in the graph topology. To complement the current section we discuss a special case where we are able to solve (2.44).

Assume that the delays to a single node are all equal, i.e. $\tau_{ki} = \tau_i$, then (2.44) gets notably simpler, i.e.

$$\Omega = \omega_i - \frac{K}{d_i} \sum_{k=1}^{N} a_{ki} \sin(\Omega \tau_i) = \omega_i - \frac{K}{d_i} \sin(\Omega \tau_i) \underbrace{\sum_{k=1}^{N} a_{ki}}_{=d_i}. \qquad (2.47)$$

Furthermore, we assume that all ω_i are positive. If Ω and K are chosen such that $\Omega < \min_{i \in \{1, \ldots, N\}} \omega_i$ and $\frac{\omega_i - \Omega}{K} < 1$ for all $i \in \{1, \ldots, N\}$, then the τ_i are found by solving

$$\tau_i = \frac{1}{\Omega} \arcsin\left(\frac{\omega_i - \Omega}{K}\right). \qquad (2.48)$$

An example where the assumption that the delays to a single node are all equal holds, is the case of a fixed directed ring graph topology. In other words, the N delay differential equations (2.36) are of the form

$$\dot{\theta}_i(t) = \omega_i + K\sin(\theta_{i-1}(t - \tau_{i,i-1}) - \theta_i(t)), \tag{2.49}$$

for $i \in \{1, \ldots, N\}$.

2.2.4 Sufficient conditions for synchronization

Preliminary result

In the present section, we give the basic statement for our reasoning. We consider a large scale system consisting of N coupled one-dimensional subsystems. Each subsystem has a state θ_i whose dynamics are given by the RFDE

$$\dot{\theta}_i(t) = \frac{K}{d_i} \sum_{k=1}^{N} a_{ki} g_{ki}(\theta_{t,1}, \ldots, \theta_{t,N}) f_{ki}(\theta_k(t - \tau_{ki}) - \theta_i(t)), \tag{2.50}$$

with state $\theta_i \in \mathbb{R}$, coupling strength $K \in \mathbb{R}$, a scaling factor $d_i = \sum_k a_{ki}$, smooth nonlinear coupling gains $g_{ki} : C_T \to \mathbb{R}$, smooth coupling functions $f_{ki} : \mathbb{R} \to \mathbb{R}$, an adjacency matrix (a_{ki}) describing the graph $G(A)$ and constant delays described by $0 \leq \tau_{ki} \in \mathbb{R}$. Our result on model (2.50) generalizes a result from Papachristodoulou et al. [107], where the case of $g_{ki} = $ const. is considered. This generalization allows the investigation of frequency synchronization for delayed models where the associated graph is directed, since models of the form (2.50) typically appear in the analysis of the frequency synchronization problem. As consequence, we are able to show frequency synchronization and phase synchronization in a unified framework.

More specifically, our preliminary statement deals with (2.50) instead of (2.36), because the dynamics (2.50) play a crucial role in the proofs for both frequency synchronization and phase synchronization. In the case of frequency synchronization for delayed models with an associated all-to-all graph, we consider the second derivative of θ_i in (2.36), which is a special version of (2.50). In the general case of frequency synchronization and the case of phase synchronization, (2.50) reappears when we consider the dynamics of (2.36) in a rotating frame with common frequency. Hence, in the following lemma, conditions are given which guarantee agreement, i.e. $\theta_i(t; \phi) = \theta^*$ for $t \to \infty$ and all $i \in \{1, \ldots, N\}$ for (2.50). We utilize this result for the subsequently following sufficient conditions for frequency and phase synchronization.

Lemma 2.22. *Assume for all $t \geq 0$ and all $k, i \in \{1, \ldots, N\}$*

$$a_{ki} \neq 0 \Rightarrow g_{ki}(\theta_{t,1}, \ldots, \theta_{t,N}) > 0. \tag{2.51}$$

Furthermore assume there exists an interval $[-\gamma, \gamma] \subset \mathbb{R}$ such that for all $y \in [-\gamma, \gamma] \setminus \{0\}$ and all $k, i \in \{1, \ldots, N\}$

$$y f_{ki}(y) > 0. \tag{2.52}$$

If the graph $G(A)$ of system (2.50) contains a spanning tree, then $\Xi = C([-T, 0], [-\frac{\gamma}{2}, \frac{\gamma}{2}]^N)$ is positively invariant with respect to (2.50) and there is a $\theta^* \in [-\frac{\gamma}{2}, \frac{\gamma}{2}]$ such that for all initial conditions $\phi \in \Xi$, i.e.

$$\forall i \in \{1, \ldots, N\} : \lim_{t \to \infty} \theta(t; \phi) = \theta^*. \tag{2.53}$$

Proof. See Section 2.2.5. ∎

Frequency synchronization

The present section establishes a result for non-identical Kuramoto oscillators with heterogeneous delays in the case of all-to-all coupling, i.e. we consider a model of the form

$$\dot{\theta}_i(t) = \omega_i + \frac{K}{N} \sum_{k=1}^{N} \sin\left(\theta_k(t - \tau_{ki}) - \theta_i(t)\right), \ i \in \{1, \dots, N\}. \tag{2.54}$$

Note that in contrast to the general model (2.36), we assume that the underlying graph is an all-to-all undirected graph and the scaling is done proportional to the in-degree of each node, i.e. $d_i = N$. Model (2.54) is the classical Kuramoto model with heterogeneous delays.

For our result, we first explain the connection between (2.54) and Lemma 2.22. Consider a solution $t \mapsto \theta(t; \phi) = (\theta_1(t; \phi), \dots, \theta_N(t; \phi))^\mathsf{T}$ of (2.54). Then $t \mapsto \ddot{\theta}(t; \phi)$ fulfills

$$\ddot{\theta}_i = -\frac{K}{N} \sum_{k=1}^{N} \cos\left(\theta_i(t) - \theta_k(t - \tau_{ki})\right)\left(\dot{\theta}_i(t) - \dot{\theta}_k(t - \tau_{ki})\right). \tag{2.55}$$

We obtain (2.50) from (2.55) by setting $g_{ki}(\theta_{t,1}, \dots, \theta_{t,N}) = \cos\left(\theta_i(t) - \theta_k(t - \tau_{ki})\right)$ and $f(y) = y$ and by the substitution of $\dot{\theta}_i$ for θ_i. In other words, (2.55) is a special case of (2.50). We see immediately that $f(y) = y$ fulfills the condition $yf(y) > 0$ for all $y \in \mathbb{R}$. The issue is to fulfill the condition $g_{ki}(\theta_{t,1}, \dots, \theta_{t,N}) = \cos\left(\theta_i(t) - \theta_k(t - \tau_{ki})\right) > 0$ for all $t \geq 0$. The next theorem gives conditions which allow us to infer frequency synchronization for (2.54).

Theorem 2.23. *Set* $\omega_{max} = \max_{i \in \{1, \dots, N\}} \omega_i$, $\omega_{min} = \min_{i \in \{1, \dots, N\}} \omega_i$ *and* $\omega_M = \max_{i \in \{1, \dots, N\}} |\omega_i|$. *Let* $Q = \left\{x \in \mathbb{R} \,\middle|\, |x| \leq \frac{\pi}{2} - \delta\right\}$ *for a* $\delta \in (0, \frac{\pi}{4})$ *and* $T = \max_{k, i \in 1, \dots, N} \tau_{ki}$. *Assume that*

a) *the initial condition* $\phi = (\phi_1, \dots, \phi_N)$ *of (2.54) satisfies* $\phi_i(0) - \phi_j(0) \in Q$ *and the rate bound* $|\dot{\phi}_i(s)| \leq K + \omega_M$ *for all* $i, j \in \{1, \dots, N\}$ *and all* $s \in [-T, 0]$ *and*

b) *the maximal delay* T *satisfies*

$$T < \min\left\{\frac{\arcsin\left(\cos(\delta) - \frac{(\omega_{max} - \omega_{min})}{K}\right)}{2(\omega_M + K)}, \frac{\delta}{2(\omega_M + K)}\right\} \tag{2.56}$$

Then the Kuramoto model (2.54) achieves frequency synchronization.

Proof. See Section 2.2.5. ■

Remark 2.24. *The rate bound* $|\dot{\theta}_i(s)| \leq K + \omega_M$ *on the initial condition in Theorem 2.23 a) is important to guarantee that the state difference does not grow beyond* $\frac{\pi}{2}$. *It is always possible to satisfy the rate bound for sufficiently large K, in other words it is always possible to choose K large enough to compensate strongly varying initial conditions. A typical case where the rate bound is satisfied automatically is the case of constant initial conditions.*

Remark 2.25. *Theorem 2.23 also yields a condition for the undelayed case. Since $T = 0$ in the undelayed case, condition (2.101) from the proof of Theorem 2.23 holds with $\eta = \cos \delta$, i.e.*

$$\dot{V} \leq \omega_{max} - \omega_{min} - N \cos \delta. \tag{2.57}$$

Therefore, choosing K such that

$$K > \frac{(\omega_{max} - \omega_L)}{\cos(\delta)}, \qquad (2.58)$$

is sufficient to achieve $\left| \theta_i(t) - \theta_j(t) \right| \le \frac{\pi}{2} - \delta$ *for all* $i, j \in \{1, \ldots, N\}$ *and* $t \ge 0$. *The bound given by the inequality (2.58) for the sufficient coupling gain to achieve the positive invariance is less conservative than the bound proposed in Theorem 4.1 in Chopra and Spong [30] for the case of the undelayed Kuramoto oscillator. More importantly, the bound given in [30] scales with the number of oscillators N. Our bound has the property that it is independent of N. Therefore, our bound from inequality (2.58) complements the analysis of [30] by giving a sufficient coupling gain for the onset of frequency synchronization which is in the same order of magnitude as the necessary bound for frequency synchronization in [30]. The results for the undelayed case are supported by the findings in Dörfler and Bullo [35].*

Frequency synchronization and phase synchronization for directed graphs

After establishing in the previous section the result for frequency synchronization if the associated graph is all-to-all, we consider now the problem of phase synchronization and frequency synchronization where the associated graphs are directed graphs and contain a directed spanning tree. Therefore we consider again the general model of heterogeneous Kuramoto oscillators with heterogeneous delays as given by (2.36). The approach we utilize in this section differs from the previous section in the assumption that we have a solution for the associated self-consistency conditions from Section 2.2.3. As discussed in Section 2.2.3, this is a restrictive assumption. First we discuss the case of phase synchronization and recall a result from Papachristodoulou et al. [107]. Then we show that Lemma 2.22 allows a reasoning for the case of frequency synchronization similar to the case of phase synchronization.

In contrast to the undelayed case, where phase synchronization cannot be achieved in a network of non-identical oscillators (see Section 2.2.3), the following section gives conditions that allow phase synchronization in the non-identical case if delays are present, i.e. there is a $(k,i) \in \{1,\ldots,N\} \times \{1,\ldots,N\}$ such that $\tau_{ki} \ne 0$. We assume in the following that there exists a solution Ω for the self-consistency condition (2.44), which is necessary for phase synchronization. Consider the dynamics of oscillator i in a rotating frame. In other words, consider the coordinate transformation defined by $\theta_i(t) = \Omega t + \xi_i(t)$. Differentiation of both sides of $\theta_i(t) = \Omega t + \xi_i(t)$ with respect to time and substitution of (2.44) results in

$$\dot{\xi}_i(t) = \frac{K}{d_i} \sum_{k=1}^{N} a_{ki} \Big(\sin(-\Omega \tau_{ki} + \xi_k(t - \tau_{ki}) - \xi_i(t)) + \sin(\Omega \tau_{ki}) \Big). \qquad (2.59)$$

Again, (2.59) is a special form of (2.50) which we obtain by setting $g_{ki} = 1$ and

$$f_{ki}(y) = \sin(-\Omega \tau_{ki} + y) - \sin(-\Omega \tau_{ki}) \qquad (2.60)$$

and substituting ξ_i for θ_i. We see directly that the condition (2.51) on g from Lemma 2.22 is fulfilled. In order to apply Lemma 2.22, we have to determine the set where $y f_{ki}(y) > 0$ for all $y \ne 0$ holds.

We know that for a fixed τ_{ki}, the function (2.60) is zero for $y = 2l\pi$ and for $y = 2\Omega \tau_{ki} - \pi - 2\pi l, l \in \mathbb{Z}$. Now, let $T = \max_{i,k} \tau_{ki}$, and assume throughout that there is an Ω which solves (2.44) such that $\cos(\Omega \tau_{ki}) > 0$ for all $i, k = \{1, \ldots, N\}$. Then the derivative of f_{ki} is positive at

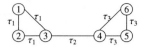

Figure 2.4: The network topology for the simulation example. The delays are chosen in order to simplify the computation of the remaining parameters of the system.

$y = 2l\pi$. Since f_{ki} is a translated sine, $y f_{ki}(y) > 0$ in intervals around $y = 2k\pi$. The size of the interval depends on $\Omega \tau_{ki}$. Therefore let γ be defined by

$$\gamma = \min_{\substack{i,k \in \{1,\dots,N\} \\ l \in \mathbb{Z}}} |2l\pi + \pi - 2\Omega \tau_{ki}|. \tag{2.61}$$

With these considerations we obtain a theorem which gives conditions for phase synchronization in delayed Kuramoto models where the associated directed graph contains a spanning tree. A similar result appeared in Papachristodoulou et al. [107, Corollary] for the case of a connected, undirected graph.

Theorem 2.26. *Consider a Kuramoto model of the form* (2.36) *where the associated directed graph contains a spanning tree. Assume that the self-consistency condition* (2.44) *has a solution* Ω *such that* $\cos \Omega \tau_{ik} > 0$ *for all* $i,k \in \{1,\dots,N\}$. *Define* $T = \max_{i,k \in \{1,\dots,N\}} \tau_{ik}$ *and* γ *as in* (2.61). *Furthermore, suppose the initial conditions* ϕ *satisfy for all* $i \in \{1,\dots,N\}$ *and all* $s \in [-T, 0]$

$$|\phi_i(s)| \le \frac{\gamma}{2}. \tag{2.62}$$

Then the Kuramoto model (2.54) *achieves phase synchronization.*

Proof. We already saw in the introduction of this section that the dynamics in the rotating frame (2.59) are a special case of (2.50) and that g from (2.50) equals one. Furthermore we defined γ in (2.61) such that $y f(y) > 0$ for all $y \in [-\frac{\gamma}{2}, \frac{\gamma}{2}] \setminus \{0\}$. From our assumption on the existence of a spanning tree in the graph associated with the model, we get attractivity of the phase synchronization set by Lemma 2.22. ∎

Remark 2.27. *One important assumption of the proof is the existence of a solution* Ω *for the self-consistency condition* (2.44). *However, it is not obvious how to give general conditions that allow to conclude the existence of a solution of the self-consistency condition* (2.44), *as already discussed in Section 2.2.3. There, we gave solution conditions under restrictive assumptions. Note that even for a given set of parameters, the problem is in general not easy to solve. However, it is possible to check for a given parameter set if* (2.44) *does not have a solution.*

Example 2.28. *To illustrate the phase synchronization effect, consider a network of Kuramoto oscillators with a topology as in Figure 2.4. The dynamics for the example are given by the standard non-identical Kuramoto equations with heterogeneous coupling, i.e.*

$$\dot{\theta}_i(t) = \omega_i + \frac{K}{d_i} \sum_{k=1}^{N} a_{ki} \sin\left(\theta_k(t - \tau_{ki}) - \theta_i(t)\right). \tag{2.63}$$

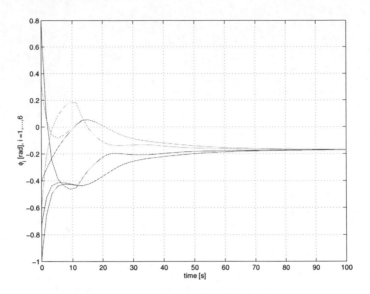

Figure 2.5: Simulation results for a Kuramoto model with delays where the associated graph has the topology from Figure 2.4 and parameters as given in Example 2.28. The figure shows exemplary solutions in the rotating frame.

In order to simplify the computation of the corresponding frequencies, we choose the coupling and the delays of the network and set the coupling strength to $K = 1$. We define $\tau_1 = \tau_3 = 1$ and $\tau_2 = 10$ and choose Ω and the ω_i according to the consistency condition (2.44). We have the following set of parameters

$$\Omega = 0.6905, \quad \omega_1 = \omega_2 = \omega_5 = \omega_6 = 0.90286, \quad \omega_3 = \omega_4 = 1.0. \tag{2.64}$$

From (2.61) we obtain a $\gamma = 1.7606$. The initial conditions were chosen as constant functions, i.e. for all $s \in [-10, 0]$ we have

$$\phi_1(s) = -0.7 \qquad \phi_2(s) = -1 \qquad \phi_3(s) = 0.8$$
$$\phi_4(s) = -0.6 \qquad \phi_5(s) = -0.4 \qquad \phi_6(s) = 0.3. \tag{2.65}$$

Figure 2.5 shows the time evolution of the phases ξ_i in the rotating frame. We achieve phase synchronization despite the fact that $|\phi_2(s)| = 1 > \frac{\gamma}{2}$ holds. Therefore, the conditions in Theorem 2.26 are only sufficient.

It is possible to carry out the transformation into the rotating frame also for the case of frequency synchronization. First, we consider the self-consistency condition for this case, as given by

$$\Omega = \omega_i + \frac{K}{d_i} \sum_{k=1}^{N} \sin\left(\Omega \tau_{ki} + \delta_k - \delta_i\right). \tag{2.66}$$

We assume there are δ_i such that the self-consistency condition has a solution Ω. Then we rewrite the time evolution of the phases to

$$\theta_i(t) = \Omega t + \xi_i(t) + \delta_i. \tag{2.67}$$

A transformation into the ξ_i-coordinate then results in

$$\dot{\xi}_i(t) = \frac{K}{d_i}\left(\sum_{k=1}^{N} a_{ki}\Big(\sin(\Omega\tau_{ki} + \delta_k - \delta_i) + \sin\big(\xi_k(t - \tau_{ki}) - \xi_i(t) + \delta_k - \delta_i - \Omega\tau_{ki}\big)\Big)\right). \tag{2.68}$$

The previous equation is not notably different from (2.59), thus the proof can be repeated for the case of frequency synchronization. Hence we immediately obtain the following result.

Corollary 2.29. *Consider a Kuramoto model of the form* (2.36) *where the associated graph contains a spanning tree. Define* $T = \max_{i,j\in\{1,...,N\}} \tau_{ij}$ *and let* γ *be defined by* (2.61). *Furthermore, suppose the initial conditions* ϕ *satisfy for all* $i \in \{1,...,N\}$ *and all* $s \in [-T,0]$

$$|\phi_i(s)| \leq \frac{\gamma}{2}. \tag{2.69}$$

Then the Kuramoto model (2.54) *achieves frequency synchronization.*

In contrast to Theorem 2.23, Corollary 2.29 works for more general topologies. However, Corollary 2.29 does not allow to determine the region of attraction directly from the model (2.36), as it was possible in the all-to-all coupling case. In the general case one would have to determine γ beforehand with the help of equation (2.42). Due to the problems concerning the solution of equation (2.42), as discussed in Remark 2.27, the frequency synchronization conditions obtained by Corollary 2.29 are more difficult to apply than the ones obtained for the all-to-all case in Theorem 2.23.

2.2.5 Technical proofs

In the proof of Lemma 2.22 and in the proofs of the subsequent lemmata and theorems, we frequently reuse the concept of a maximal state for (2.36). To formalize the use in the rest of the section, we use a notation which is summarized in the following remark.

Remark 2.30. *For* $i \in \{1,...,N\}$, *let* $\theta_i : \mathbb{R} \to \mathbb{R}$ *denote the solutions of* (2.50). *By* $\theta_U(t;\phi) = \max_{i\in\{1,...,N\}} \theta_i(t;\phi)$ *we define the function which attains the maximal value over all solutions at each time point t. If there are several possible indices to choose, then* $\theta_U(t;\phi)$ *is the* $\theta_i(t;\phi)$ *with the maximal derivative at t. If there are still several possibilities, we choose one of the candidate and fix it as long as it satisfies the maximum conditions. We define the function* $\theta_L(t;\phi) = \min_{j\in\{1,...,N\}} \theta_j(t;\phi)$ *in an analogous fashion.*

The proof of Lemma 2.22

The proof of Lemma 2.22 is along the lines of Papachristodoulou et al. [107, Section II]. Similar to Papachristodoulou et al. [107, Section II], our proof is mainly based on the invariance principle for RFDEs from Haddock and Terjeki [52]. Our goal is to prove attractivity of subspaces of C_T.

Let D be a subset of $\mathbb{R}^n$ and denote by $V : D \to \mathbb{R}$ a continuous function. The derivative of V along solutions of an RFDE (2.39) is defined by

$$\dot{V}(\phi) = \limsup_{h \to 0^+} \frac{1}{h} \Big(V\big(\phi(0) + hf(\phi)\big) - V(\phi(0)) \Big). \tag{2.70}$$

For a given set $\Xi \subset C_T$, define

$$E_V = \Big\{ \phi \in \Xi \, \big| \, \forall t \geq 0 : \max_{s \in [-T,0]} V\big(x_t(s;\phi)\big) = \max_{s \in [-T,0]} V\big(\phi(s)\big) \Big\}, \tag{2.71}$$

and denote by $M_V \subset E_V$ the largest positively invariant set with respect to (2.39). Here, E_V is the set of functions $\phi \in \Xi$ which can serve as initial conditions for (2.39) such that $x_t(\phi)$ satisfies $\max_{s \in [-T,0]} V\big(x_t(\phi)(s)\big) = \max_{s \in [-T,0]} V\big(\phi(s)\big)$ for all $t \geq 0$. Then the following holds:

Lemma 2.31. *(Haddock and Terjeki [52, Theorem 2.1]) Suppose there exists a function V and a closed set $\Theta \subset C_T$ that is positively invariant with respect to (2.39). Furthermore assume that for all $\phi \in \Theta$ such that $V(\phi(0)) = \max_{s \in [-T,0]} V(\phi(s))$ the derivative of V with respect to $\dot{x} = f(x_t)$ fulfills*

$$\dot{V}(\phi) \leq 0. \tag{2.72}$$

Then, for any $\phi \in \Theta$ such that $t \mapsto x(t;\phi)$ is defined and bounded on $[-T, \infty)$, $\omega(\phi) \subseteq M_V \subseteq E_V$, and we have $x_t(\phi) \to M_V$ for $t \to \infty$.

Note that we consider throughout this section the rate of change of functions along solutions of (2.39), i.e. we use arguments of Razumikhin type. As consequence we refer to functions of which we consider the rate of change along solutions of RFDEs as *Lyapunov-Razumikhin* functions.

Proof. The proof of Lemma 2.22 is an application of Lemma 2.31. As consequence we divide the proof in three steps. First, we are going to use the condition (2.52) to establish positive invariance of $C([-T,0], [-\frac{\gamma}{2}, \frac{\gamma}{2}]^N)$. With the help of the positive invariance, we show in the second step that there is a Lyapunov Razumikhin function V such that its derivative is non-positive at the points of concern. The final step shows that the existence of a spanning tree guarantees the attractivity of the phase synchronization set.

First, we show the positive invariance of $C([-T,0], [-\frac{\gamma}{2}, \frac{\gamma}{2}]^N)$. Therefore consider a initial conditions ϕ_i which fulfills

$$-\frac{\gamma}{2} \leq \phi_i(s) \leq \frac{\gamma}{2} \tag{2.73}$$

for $s \in [-T,0]$. Suppose that the positive invariance of $C([-T,0], [-\frac{\gamma}{2}, \frac{\gamma}{2}]^N)$ does not hold, i.e. $t \mapsto (\theta_1(t;\phi), \ldots, \theta_N(t;\phi))^\top$ leaves the set $[-\frac{\gamma}{2}, \frac{\gamma}{2}]^N$. Then, there is a smallest $t^* > 0$ such that

a) for all $s \in [-T, t^*)$ and for all $i \in \{1, \ldots, N\}$ we have $\frac{\gamma}{2} \leq \theta_i(s;\phi) \leq \frac{\gamma}{2}$, $\dot{\theta}_i(s;\phi) \leq 0$ if $\theta_i(s;\phi) = \frac{\gamma}{2}$ and $\dot{\theta}_i(s;\phi) \geq 0$ if $\theta_i(s) = -\frac{\gamma}{2}$,

b) at t^* there is an $i \in \{1, \ldots, N\}$ such that we have either $\theta_i(t^*;\phi) = \frac{\gamma}{2}$ and $\dot{\theta}_i(t^*;\phi) > 0$ or $\theta_i(t^*;\phi) = -\frac{\gamma}{2}$ and $\dot{\theta}_i(t^*;\phi) < 0$.

Suppose the case where $\theta_i(t^*;\phi) = \frac{\gamma}{2}$ and $\dot{\theta}_i(t^*;\phi) > 0$ holds. Then $\theta_i(t^*;\phi) = \frac{\gamma}{2} \geq \theta_k(t^* - \tau_{ki};\phi)$ and $|\theta_k(s;\phi)| \leq \frac{\gamma}{2}$ for $s \in [-T, t^*)$ and all $k \in \{1, \ldots, N\}$. The dynamics for θ_i are determined by

$$\dot{\theta}_i(t) = \frac{K}{d_i} \sum_{k=1}^N a_{ki} g_{ki}\big(\theta_{1,t}, \ldots, \theta_{N,t}\big) f_{ki}\big(\theta_k(t - \tau_{ki}) - \theta_i(t)\big), \tag{2.74}$$

where f_{ki} satisfies (2.52).

$\theta_i(t^*; \phi) \geq \theta_k(t^* - \tau_{ki}; \phi)$ implies $\theta_k(t^* - \tau_{ki}; \phi) - \theta_i(t^*; \phi) \leq 0$ and since $yf_{ki}(y) > 0$, we know $f_{ki}(\theta_k(t^* - \tau_{ki}; \phi) - \theta_i(t^*; \phi)) \leq 0$. As consequence the inequality

$$\dot{\theta}_i(t^*; \phi) \leq 0 \tag{2.75}$$

holds, which is a contradiction to the assumption $\dot{\theta}_i(t^*; \phi) > 0$. The assumption $\dot{\theta}_i(t^*; \phi) < 0$ for $\theta_i(t^*; \phi)$ leads to a similar contradiction. Overall, we conclude that

$$-\frac{\gamma}{2} \leq \theta_i(t; \phi) \leq \frac{\gamma}{2} \tag{2.76}$$

for all $t \geq -T$ and all $i \in \{1, \ldots, N\}$. In other words $C([-T, 0], [-\frac{\gamma}{2}, \frac{\gamma}{2}]^N)$ is positively invariant.

For the second step we show the existence of Lyapunov-Razumikhin function candidates that have non-positive derivatives along the solutions of (2.74) on $C([-T, 0], [-\frac{\gamma}{2}, \frac{\gamma}{2}]^N)$. Therefore consider

$$V_1 = \max_{i \in \{1, \ldots, N\}} \theta_i, \quad V_2 = -\min_{i \in \{1, \ldots, N\}} \theta_i. \tag{2.77}$$

In the following we utilize the notation from Remark 2.30, i.e. $\theta_U(t; \phi) = \max_{i \in \{1, \ldots, N\}} \theta_i(t; \phi)$ and $\theta_L(t; \phi) = \min_{i \in \{1, \ldots, N\}} \theta_i(t; \phi)$. Calculating the derivatives of V_1 and V_2 along solutions of (2.50) results in

$$\dot{V}_1 = K \sum_{k=1}^{N} \frac{a_{kU}}{d_U} g_{kU}(\theta_{1,t}, \ldots, \theta_{N,t}) f_{kU}(\theta_k(t - \tau_{kU}) - \theta_U(t)), \tag{2.78}$$

$$\dot{V}_2 = -K \sum_{k=1}^{N} \frac{a_{kL}}{d_L} g_{kL}(\theta_{1,t}, \ldots, \theta_{N,t}) f_{kL}(\theta_k(t - \tau_{kL}) - \theta_L(t)). \tag{2.79}$$

According to Lemma 2.31 we have to consider $\dot{V}_1$ and $\dot{V}_2$ at the points where $V_i(\theta(t)) = \max_{s \in [-T,0]} V_i(\theta(t+s))$, $i \in \{1, 2\}$. Thus, by construction of V_i, we have to consider $\dot{V}_i$ at

$$\theta_U = \max_{s \in [-T,0]} \max_{i \in \{1, \ldots, N\}} \theta_i(t+s) \quad \text{and} \quad \theta_L = \min_{s \in [-T,0]} \min_{i \in \{1, \ldots, N\}} \theta_j(t+s). \tag{2.80}$$

Property (2.51) for g and property (2.52) for f together with $\theta_k(t - \tau_{kU}) - \theta_U \leq 0$ and $\theta_k(t - \tau_{kU}) - \theta_L \geq 0$ guarantee $\dot{V}_i \leq 0$, $(i \in \{1, 2\})$ at these points.

In the remaining part of the proof we determine M_V, which determines the asymptotic dynamics of (2.36) with initial conditions in the set Ξ which is given by $\Xi = C([-T, 0], [-\frac{\gamma}{2}, \frac{\gamma}{2}]^N) \subset C_T$. By Lemma 2.31, M_V is the largest set in E_{V_i} that is invariant with respect to (2.50), where E_{V_i} is a subset of the positively invariant set Ξ. First, determine the sets E_{V_i} defined in (2.71) for Ξ, i.e.

$$E_{V_i} = \left\{ \phi \in \Xi \middle| \forall t \geq 0 : \max_{s \in [-T,0]} V_i(\theta_t(s; \phi)) = \max_{s \in [-T,0]} V_i(\phi(s)) \right\}. \tag{2.81}$$

In a similar way as explained in Haddock and Terjeki [52, Remark 1.1], we utilize here that for any $\phi \in E_{V_i}$ the condition $\dot{V}_i(\theta_t(s; \phi)) = 0$ holds for any $t > 0$ where

$$\max_{s \in [-T,0]} V_i(\theta_t(s; \phi)) = V_i(\theta_t(0; \phi)) \tag{2.82}$$

holds. Now consider E_{V_i} in the case $i = 1$ and let θ_1^* be given by $\theta_1^* = \max_{s \in [-T,0]} V_1(\phi(s))$. In this situation equation (2.82) is $\max_{s \in [-T,0]} \theta_{U,t}(s; \phi) = \theta_{U,t}(0; \phi) = \theta_U(t; \phi)$. From the

35

definition of E_{V_1} we consider the $\phi \in \Xi$ where $\max_{s \in [-T,0]} \theta_{U,t}(s; \phi) = \max_{s \in [-T,0]} V_1(\phi(s)) = \theta_1^*$. Combining the two previous statements, we have $\dot{\theta}_U(t; \phi) = 0$ where $\theta_U(t; \phi) = \theta_1^*$.

The derivative $\dot{V}_1$ from (2.78) together with the condition that g is positive from property (2.51), the property (2.52) for f and the equation $\dot{V}_1(\theta_t(0; \phi)) = \dot{\theta}_U = 0$ imply

$$\forall k : e_{kU} \in \mathscr{E} : \theta_k(t - \tau_{kU}; \phi) = \theta_U(t; \phi) = \theta_1^*. \tag{2.83}$$

In other words, the phase θ_k of oscillator k fulfills $\theta_k(t - \tau_{kU}; \phi) = \theta_1^*$ if oscillator k is a parent of oscillator U. Hence for all parents k of oscillator U we have $\dot{\theta}_k(t - \tau_{kU}; \phi) = 0$, since $\theta_k(t - \tau_{kU}; \phi) = \theta_1^*$ and we know that $\theta_1^* = \max_{s \in [-T,0]} V_1(\phi(s)) = \max_{s \in [-T,0]} V_1(\theta_{t-\tau_{kU}}(\phi)(s))$, see (2.71). As consequence, we continue the argument to the parents of the parents etc., until we reach any root of the underlying graph $G(A)$. By assumption, the graph contains a spanning tree, i.e. at least one root. Therefore, the initial condition of all roots in the graph need to take the value θ_1^* at some $t \in [-T, 0]$. Consequently, E_{V_1} is given as:

$$E_{V_1} \subset \bigcup_{\theta_1^* \in \mathbb{R}} \left\{ \phi \in \Xi | \forall i \in V(G)_R | \max_{s \in [-T,0]} \phi_i(s) = \theta_1^*, \right.$$
$$\left. \forall i \in V(G) \setminus V(G)_R : \forall s \in [-T,0] : \phi_i(s) \le \theta_1^* \right\}. \tag{2.84}$$

We obtain a similar condition for E_{V_2}, i.e.

$$E_{V_2} \subset \bigcup_{\theta_2^* \in \mathbb{R}} \left\{ \phi \in \Xi | \forall i \in V(G)_R | \min_{s \in [-T,0]} \phi_i(s) = \theta_2^* \right.$$
$$\left. \forall i \in V(G) \setminus V(G)_R : \forall s \in [-T,0] : \phi_i(s) \ge \theta_2^* \right\}. \tag{2.85}$$

with $\theta_2^* = \min_{s \in [0,T]} \min_{i \in \{1,...,N\}} \theta_i(t - s)$ for all $t \ge 0$.

For all $\phi \in E_{V_1} \cap E_{V_2}$, we have $\max_{s \in [-T,0]} \theta_{i,t}(s; \phi) = \theta_1^*$ and $\min_{s \in [-T,0]} \theta_{i,t}(s; \phi) = \theta_2^*$ for all $t \ge 0$ and all $i \in V(G)_R$ and all solutions $t \mapsto (\theta_1(t; \phi), \dots, \theta_1(t; \phi))^\mathsf{T}$ of (2.74) with initial condition ϕ. Hence there are two exclusive possibilities for ϕ:

a) For all $s \in [-T, 0]$ the equation $\phi_i(s) = \theta_1^* = \theta_2^*$ holds, or

b) ϕ leads to a persistently oscillating solution $t \mapsto \theta(t; \phi)$.

We show with a brief reasoning why b) cannot hold for (2.74). Consider i and t such that $\theta_i(t; \phi) = \theta_1^*$ and remember that $\theta_{k,t}(s; \phi) \in [\theta_2^*, \theta_1^*]$ for all $t \ge 0$ and for all k. For a persistently oscillating solution, we would need that one $t \mapsto \theta_i(t; \phi)$ reaches $\theta_2^* < \theta_1^*$ in finite time. Observe that

$$\dot{\theta}_i(t) = \frac{K}{d_i} \sum_{k=1}^{N} a_{ki} g_{ki}(\theta_{1,t}, \dots, \theta_{N,t}) f_{ki}(\theta_k(t - \tau_{ki}) - \theta_i(t)) \tag{2.86}$$

$$\ge K_i \max_k g_{ki}(\theta_{1,t}, \dots, \theta_{N,t}) f_{ki}(\theta_2^* - \theta_i) \ge \alpha(\theta_2^* - \theta_i). \tag{2.87}$$

Since g_{ki} is continuous and the argument of g_{ki} is bounded for $T < \infty$ and since we can bound f_{ki} on a bounded interval by a linear function because $x f_{ki}(x) > 0$, we always find a finite $\alpha \in \mathbb{R}$ that satisfies the previous inequality. Hence we see that θ_i approaches θ_2^* not faster than exponentially, especially not in finite time.

Because of Lemma 2.31 any $\phi \in \Xi$ such that $\theta_t(s; \phi)$ is defined and bounded on $[-T, \infty]$ fulfills $\omega(\phi) \subseteq M_V \subseteq E_V$. Because both Lyapunov-Razumikhin functions V_1 and V_2 fulfill the conditions, we get $\omega(\phi) \subseteq M_V \subseteq E_{V_1}$ and $\omega(\phi) \subseteq M_V \subseteq E_{V_2}$. Together, we thus obtain $\omega(\phi) \subseteq M_V \subseteq E_{V_1} \cap E_{V_2}$. Since $\omega(\phi)$ is nonempty, see Hale and Lunel [55, Chapter 4], also M_V is nonempty. In other words, all elements of M_V are characterized by the elements of $E_{V_1} \cap E_{V_2}$.

If we consider now any $\phi \in E_{V_1} \cap E_{V_2}$, then $\theta_1^* = \phi_i(s) = \theta_2^*$ for all $s \in [-T, 0]$ if $i \in V(G)_R$, i.e. i is a root. If i is not a root, we directly obtain $\theta_2^* \leq \phi_i(s) \leq \theta_1^* = \theta_2^*$. Because for all $\phi \in \Xi$, $\theta_t(\phi) \to \omega(\phi) \subseteq M_V \subseteq E_{V_1} \cap E_{V_2}$ for $t \to \infty$, we therefore get $\lim_{t \to \infty} \theta_i(t; \phi) = \theta^*$ for a $\theta^* \in \mathbb{R}$ and for all $i \in \{1, \ldots, N\}$. ∎

The proof of Theorem 2.23

Proof. The proof consists of two parts. In the first part, we show that the conditions of the theorem are sufficient to guarantee that $\left|\theta_i(t; \phi) - \theta_j(t; \phi)\right| \leq \frac{\pi}{2} - \delta$ for $t \geq 0$ as long as the difference of $\theta_i(0; \phi)$ and $\theta_j(0; \phi)$ is in $Q = \left\{x \in \mathbb{R} \big| |x| \leq \frac{\pi}{2} - \delta\right\}$ for all $i, j \in \{1, \ldots, N\}$ and the rate bound $\dot{\theta}_i(s; \phi)$ holds for all $s \in [-T, 0]$ and for all $i \in \{1, \ldots, N\}$. In the second part we show that for all $t \geq 0$ the condition $|\theta_i(t; \phi) - \theta_k(t - \tau_{ki}; \phi)| < \frac{\pi}{2}$ holds for all $i, k \in \{1, \ldots, N\}$, which allows us to conclude frequency synchronization with Lemma 2.22.

We show that $\left|\theta_i(t; \phi) - \theta_j(t; \phi)\right| \leq \frac{\pi}{2} - \delta$ for all $t \geq 0$ as long as $\left|\theta_i(0; \phi) - \theta_j(0; \phi)\right| \leq \frac{\pi}{2} - \delta$. Define V by

$$V(\theta_1, \ldots, \theta_N) = \max_{i \in \{1, \ldots, N\}} \theta_i - \min_{j \in \{1, \ldots, N\}} \theta_j. \tag{2.88}$$

and consider $\dot{V}$ at the points where $V = \frac{\pi}{2} - \delta$. We show that if that is the case, V is strictly decreasing with respect to time. Functions of the form $\max \theta_i - \min \theta_j$ are a natural choice for delay systems Haddock and Terjeki [52] and were also suggested in Moreau [99]. As long as $\dot{V} < 0$ holds at these points, we are sure that $\left|\theta_i(t; \phi) - \theta_j(t; \phi)\right|$ cannot grow beyond $\frac{\pi}{2} - \delta$. We utilize the abbreviations $\theta_U(t; \phi) = \max_{i \in \{1, \ldots, N\}} \theta_i(t; \phi)$ and $\theta_L(t; \phi) = \max_{i \in \{1, \ldots, N\}} \theta_i(t; \phi)$ as explained in Remark 2.30 and calculate $\dot{V}$ along solutions of (2.54). Therefore, we get

$$\dot{V}(\theta_1(t; \phi), \ldots, \theta_N(t; \phi)) = \dot{\theta}_U - \dot{\theta}_L$$
$$= \omega_U - \omega_L - \frac{K}{N} \sum_{k=1}^{N} \sin\left(\theta_U(t; \phi) - \theta_k(t - \tau_{kU})\right) + \sin\left(\theta_k(t - \tau_{kL}) - \theta_L(t; \phi)\right), \tag{2.89}$$

where we keep the dependence on t for the sake of clarity. In the next step we, we give estimates for the delayed states in (2.89). Therefore, we rewrite $\theta(t - \tau_{ki}; \phi)$ as

$$\theta_k(t - \tau_{ki}; \phi) = \theta_k(t; \phi) - \underbrace{\int_{-\tau_{ki}}^{0} \dot{\theta}_k(t + s; \phi) \, ds}_{\varepsilon_{ki}=}. \tag{2.90}$$

Utilizing (2.90), (2.89) results in

$$\dot{V}(\theta_1, \ldots, \theta_N) = \omega_U - \omega_L - \frac{K}{N} \sum_{k=1}^{N} \left(\sin(\theta_U - \theta_k + \varepsilon_{kU}) + \sin(\theta_k - \theta_L - \varepsilon_{kL}) \right). \tag{2.91}$$

As stated before, whenever $\theta_U - \theta_L = \frac{\pi}{2} - \delta$, we need $\dot{V}(\theta_1, \ldots, \theta_N) < 0$.

To achieve $\dot{V}(\theta_1, \ldots, \theta_N) < 0$, the term multiplied by K has to dominate the difference $\omega_{max} - \omega_{min}$. Therefore we have to find a strictly positive lower bound for

$$\sin(\theta_U - \theta_k + \varepsilon_{kU}) + \sin(\theta_k - \theta_L - \varepsilon_{kL}) \tag{2.92}$$

37

under the condition that $\theta_U - \theta_L = \frac{\pi}{2} - \delta$. Since $\sin(x) + \sin(y) = 2\sin\left(\frac{x+y}{2}\right)\cos\left(\frac{x-y}{2}\right)$, we rewrite (2.92) as

$$2\sin\left(\frac{\theta_U - \theta_L + \varepsilon_{kU} - \varepsilon_{kL}}{2}\right)\cos\left(\frac{\theta_U - \theta_k + \varepsilon_{kU} - (\theta_k - \theta_L - \varepsilon_{kL})}{2}\right). \tag{2.93}$$

Next we want to find a lower bound for the cosine term in (2.93). First, we use the relations

$$0 \leq \theta_k - \theta_L \leq \frac{\pi}{2} - \delta, \quad 0 \leq \theta_U - \theta_k \leq \frac{\pi}{2} - \delta, \tag{2.94}$$

which result directly from the definitions of θ_U and θ_L and the considered set Q. Furthermore, we find a common upper bound $\tilde{\delta}$ for $|\varepsilon_{kU}|$ and $|\varepsilon_{kL}|$ by using (2.90) and (2.54):

$$|\varepsilon_{ki}| \leq \int_{-\tau_{ki}}^{0} \left|\dot{\theta}_i(t+s)\right| ds \leq \underbrace{\max_{k,i \in \{1,\dots,N\}} \tau_{ki}}_{=T}(\omega_M + K) =: \tilde{\delta}. \tag{2.95}$$

The second term in (2.56) implies that $\tilde{\delta} < \frac{1}{2}\delta$. Therefore,

$$|\varepsilon_{kU}| \leq \tilde{\delta} < \frac{1}{2}\delta, \quad |\varepsilon_{kL}| \leq \tilde{\delta} < \frac{1}{2}\delta. \tag{2.96}$$

By summing up the results from (2.94) and (2.96) we obtain

$$-2\tilde{\delta} + \delta - \frac{\pi}{2} \leq \theta_U - \theta_k + \varepsilon_{kU} - (\theta_k - \theta_L - \varepsilon_{kL}) \leq \frac{\pi}{2} - \delta + 2\tilde{\delta}. \tag{2.97}$$

Because of the symmetry of the cosine, (2.97) directly yields to a lower bound for the cosine-term in (2.93), i.e.

$$\cos\left(\frac{\theta_U - \theta_k + \varepsilon_{kU} - (\theta_k - \theta_L - \varepsilon_{kL})}{2}\right) \geq \cos\left(\frac{\frac{\pi}{2} - \delta + 2\tilde{\delta}}{2}\right). \tag{2.98}$$

For the sine term in (2.93), the lower bound results from the fact that we consider $\dot{V}$ at the points where $\theta_U - \theta_L = \frac{\pi}{2} - \delta$ and the common bound $\tilde{\delta}$ for $|\varepsilon_{kL}|$ and $|\varepsilon_{kU}|$ from (2.96). Altogether we obtain

$$\sin\left(\frac{\theta_U - \theta_L + \varepsilon_{kU} - \varepsilon_{kL}}{2}\right) \geq \sin\left(\frac{\frac{\pi}{2} - \delta - 2\tilde{\delta}}{2}\right). \tag{2.99}$$

The inequalities (2.98) and (2.99) give a bound for (2.92), i.e.

$$\sin(\theta_U - \theta_k + \varepsilon_{kU}) + \sin(\theta_k - \theta_L - \varepsilon_{kL}) \geq$$
$$2\sin\left(\frac{\frac{\pi}{2} - \delta - 2\tilde{\delta}}{2}\right)\cos\left(\frac{\frac{\pi}{2} - \delta + 2\tilde{\delta}}{2}\right) = \cos(\delta) - \sin(2\tilde{\delta}), \tag{2.100}$$

where the last simplification results from the trigonometric relations $2\sin(x)\cos(y) = \sin(x - y) + \sin(x+y)$ and $\sin(\frac{\pi}{2} - x) = \cos(x)$. Since $\tilde{\delta} < \frac{1}{2}\delta < \frac{\pi}{8}$ we obtain $0 < \eta = \cos(\delta) - \sin(2\tilde{\delta})$. Combining (2.100) and (2.91) results in an upper bound for $\dot{V}$, i.e.

$$\dot{V}(\theta_1, \dots, \theta_N) \leq \omega_U - \omega_L - \frac{K}{N}N\eta \tag{2.101}$$
$$\leq \omega_{\max} - \omega_{\min} - K\eta.$$

As explained in the problem statement in Section 2.2.2, we consider no delay in the self-coupling, i.e. $\tau_{ii} = 0$. The factor N in (2.101) vanishes since one of the sine summands in (2.91) vanishes for $k \in \{U, L\}$, and it is possible to choose η as lower bound for the other summand. The inequality (2.101) directly yields a condition for K such that $\dot{V} < 0$ if $V = \frac{\pi}{2} - \delta$, namely

$$K > \frac{(\omega_{\max} - \omega_{\min})}{\eta}. \tag{2.102}$$

The substitutions $\eta = \cos(\delta) - \sin(2\tilde{\delta})$ and $\tilde{\delta} = T(\omega_M + K)$ in (2.102) and solving the inequality for T gives

$$T < \frac{\arcsin\left(\cos(\delta) - \frac{(\omega_{\max} - \omega_{\min})}{K}\right)}{2(\omega_M + K)}. \tag{2.103}$$

Since (2.103) holds by (2.56), we get $\dot{V} < 0$ whenever $\theta_U(t; \phi) - \theta_L(t; \phi) = \frac{\pi}{2} - \delta$, therefore $|\theta_i(t) - \theta_j(t)| \leq \frac{\pi}{2} - \delta$ for all $t \geq 0$.

We now show $|\theta_i(t; \phi) - \theta_k(t - \tau_{ki}; \phi)| < \frac{\pi}{2}$ for all $t \geq 0$. This implies $g_{ki}(\theta_{t,1}, \ldots, \theta_{t,N}) = \cos(\theta_i(t) - \theta_k(t - \tau_{ki})) > 0$ for all $t \geq 0$.

From the first part of the proof, we know that the set $Q = \{x \in \mathbb{R} \,|\, |x|_\infty \leq \frac{\pi}{2} - \delta\}$ has the property that $\theta_i(t; \phi) - \theta_k(t; \phi) \in Q$ for $i, k \in \{1, \ldots, N\}$ and $t \geq 0$. With the help of the set Q and the integration expansion for the delayed state $\theta_k(t - \tau_{ik}; \phi)$ we find the following upper bound:

$$
\begin{aligned}
|\theta_i(t; \phi) - \theta_k(t - \tau_{ik}; \phi)| &\leq \max_{i,k \in \{1,\ldots,N\}} |\theta_i(t; \phi) - \theta_k(t - \tau_{ik}; \phi)| \\
&= \max_{i,k \in \{1,\ldots,N\}} \left| \theta_i(t; \phi) - \left(\theta_k(t; \phi) - \int_{-\tau_{ik}}^{0} \dot{\theta}_k(t + s; \phi) \, ds \right) \right| \\
&\leq \max_{i,k \in \{1,\ldots,N\}} \left(|\theta_i(t; \phi) - \theta_k(t; \phi)| + \int_{-\tau_{ik}}^{0} |\dot{\theta}_k(t + s; \phi)| \, ds \right) \\
&\leq \frac{\pi}{2} - \delta + T \left(\max_{i \in \{1,\ldots,N\}} |\omega_i| + K \right).
\end{aligned}
\tag{2.104}
$$

Since (2.56) holds, we get

$$|\theta_i(t; \phi) - \theta_k(t - \tau_{ik}; \phi)| \leq \frac{\pi}{2} - \delta + T|\omega_M + K| < \frac{\pi}{2}. \tag{2.105}$$

We now collect all the facts: we know that $\theta_i(t; \phi) - \theta_k(t - \tau_{ik}; \phi) < \frac{\pi}{2}$ for all $i, k \in \{1, \ldots, N\}$ and for all $t \geq 0$, hence $\cos(\theta_i(t; \phi) - \theta_k(t - \tau_{ik}; \phi)) > 0$ for all $t \geq 0$. Furthermore we have all to all coupling, which implies the existence of a spanning tree. Hence, Lemma 2.22 implies the assertion. ∎

2.2.6 Summary

In this section, we analyzed the influence of delays in the coupling on frequency synchronization and phase synchronization for networks of Kuramoto oscillators with non-identical natural frequencies.

In case of frequency synchronization, we specifically considered the case of Kuramoto oscillators with all-to-all coupling. For this case we gave a delay-dependent lower bound on the coupling gain which is sufficient for frequency synchronization as long as the initial conditions of the system are within the specified bounds. The bounds for the coupling gain and the region of attraction depend explicitly on model parameters, i.e. the size of the Kuramoto oscillator network, the delays and the natural frequencies of the individual systems. Our analysis also improves the bounds on the coupling gain for non-delayed networks.

In addition to that, we investigate phase synchronization and frequency synchronization in delayed Kuramoto models where the associated graph is a directed graph which contains a directed spanning tree. The features of this analysis are different from the analysis for Kuramoto oscillators with all-to-all coupling. We reasoned why phase synchronization is impossible in Kuramoto oscillator networks with non-identical frequencies in the absence of delays, but showed that non-identical Kuramoto oscillators with delays may achieve phase synchronization for the case of undirected networks. A central point in this analysis is an algebraic self-consistency condition which relates the natural frequencies and the agreement frequency. In the case of phase synchronization, the absence of delays implies that the self-consistency condition only has a solution if all natural frequencies are identical. As soon as delays are present, the self-consistency condition can have a solution even though the natural frequencies are non-identical. In addition to these results, we show that the existence of a solution to the self-consistency condition for phase synchronization implies phase synchronization for certain initial conditions. A similar result holds for the case of frequency synchronization. Note that these results are not constructive, since it is not trivial to show the existence of a solution to the self-consistency condition. We complemented these theoretical results with a simulation example for the case of phase synchronization.

The self-consistency conditions are of special interest from a control theoretic point of view, since they suggest a method to control phase synchronization in a network of Kuramoto oscillators by modifying the communication delays. More specific, it is reasonable to assume that we can increase communication delays in a controlled network; if these adjustments are done in the right fashion, we would be able to enforce phase synchronization.

2.3 Synchronization in oscillator networks - a Lyapunov based approach

In this section, we investigate synchronization in an oscillator network in the sense of Definition 2.11.

As explained in Section 2.1 in the Paragraph *Lyapunov-based approach*, a common approach to synchronization in an oscillator network of N identical oscillators is to find couplings which render the zero error manifold attractive and invariant. In this section, we take this approach as a starting point for our investigations of synchronization in oscillator networks. The contributions of this section are twofold. The first contribution is an explanation that a vanishing error between the oscillator states is only necessary and not sufficient for synchronization of oscillators. We identify two conditions which together characterize synchronization. The second contribution is the presentation of an oscillator class where we can guarantee the two conditions and thus guarantee synchronization. A preliminary version of the contributions presented in this Section were published in Schmidt et al. [119].

The remainder of this section is structured as follows. We start with a problem statement in Section 2.3.1. Then we give a short literature review and some preliminaries in Section 2.3.2. In Section 2.3.3 we present our results. We illustrate the results on a simple example in Section 2.3.4. Finally, we summarize Section 2.3 in Section 2.3.5.

2.3.1 Problem statement

In this section we want to study oscillator networks for a specific class of oscillators, which is given in the following Definition.

Definition 2.32. *A Lyapunov oscillator is an oscillator*

$$\dot{x} = f(x,u), \tag{2.106}$$

such that for $u = 0$ the system (2.106) has

a) *a unique asymptotically orbitally stable periodic orbit Γ with the asymptotic phase property,*

b) *a finite set F of isolated fixed points such that the linearization at the fixed points has only eigenvalues with positive real part and*

c) *a C^1-Lyapunov function $W : \mathbb{R}^n \to \mathbb{R}$ with*

  i) *$W|_\Gamma = 0$ and $W|_{\mathbb{R}^n \setminus \Gamma} > 0$,*

  ii) *$\dot{W}(x) = \frac{\partial W}{\partial x}(x)f(x,0) = 0$ for $x \in \Gamma \cup F$ and $\dot{W}(x) = \frac{\partial W}{\partial x}(x)f(x,0) < 0$ for $x \in \mathbb{R}^n \setminus \Gamma \cup F$.*

Remark 2.33. *The properties of the function W in Definition 2.32 already imply that the points in F are unstable. If $U_\varepsilon(x_i)$ denotes an ε-neighborhood of x_i, we know for an $x_i \in F$ that $\dot{W}_i(x) < 0$, $\left(x \in U_\varepsilon(x_i) \setminus \{x_i\} \right)$, hence there are always points y in $U_\varepsilon(x_i)$ such that $W_i(y) < W_i(x_i)$ and therefore the points in F are unstable.*

Our assumption of global stability of the periodic orbit excluding a finite number of fixed points F is due to simplicity of presentation, i.e. it is possible to reformulate the results in a local setting.

We consider networks of Lyapunov oscillators of the form

$$\dot{x}_k = f(x_k, u_k), \tag{2.107}$$

where $k \in \{1,\ldots,N\}$ and where we assume that the couplings render the manifold $\mathscr{S}$ as defined in (2.35) attractive and invariant. In other words, there is a subset D of the state space of N coupled oscillators given by $\mathbb{R}^{nN}$ such that for all $k, i \in \{1,\ldots,N\}$ and all $x_0 \in D$ we have $x_k(t;x_0) - x_i(t;x_0) \to 0$ for $t \to \infty$. However, this property does in general not imply synchronization in the sense of Definition 2.13, as we discuss in detail in Section 2.3.3. Our goal is to give conditions that guarantee synchronization for networks of Lyapunov oscillators.

Problem statement 2.34. *Consider a network of Lyapunov oscillators and assume that the coupling renders $\mathscr{S}$ attractive and invariant. Find sufficient conditions for synchronization in the sense of Definition 2.13.*

In Section 2.3.3, we give equivalent conditions for synchronization in general oscillator networks under the condition that the coupling renders $\mathscr{S}$ invariant and attractive and we discuss these conditions in detail. Furthermore, we show that the conditions we derived can be guaranteed for networks of Lyapunov oscillators. Before presenting the results we give a short literature review and some preliminaries in Section 2.3.2.

2.3.2 Preliminaries

In this section, we first give a brief review of the relevant literature and then a brief account of some additional technical preliminaries.

As stated in Section 2.3.1, the main assumption that we consider here in the investigation of synchronization of oscillators is the stability and invariance of the synchronization manifold. This assumption is standard in many seminal contributions to the synchronization literature, e.g. Fujisaka and Yamada [47] or Pecora and Carroll [110]. Additional references and a more detailed overview are given e.g. in Boccaletti et al. [19, Section 5]. A common feature of these results is that they depend on a linearization of the dynamics around the synchronization manifold, see especially Pecora and Carroll [110]. In recent years, several different approaches were chosen to extend these analysis results to obtain global results. These studies were carried out under restrictive assumptions on the considered vector field. One often utilized assumption is a Lipschitz condition for the vector fields, see e.g. Li and Chen [83]. Another assumption is the so called contracting assumption, see e.g. Slotine et al. [136]. For a recent discussion and a generalization of these assumptions see DeLellis et al. [34]. These assumptions on the vector fields imply the (global) Lyapunov stability of a single solution in the system and are thus rather restrictive. Furthermore, many of these studies do not consider the asymptotic solutions in detail and just define the attractivity and invariance of the synchronization manifold as synchronization. We show in Section 2.3.3 that this is merely a necessary and no sufficient condition for synchronization of oscillators. Two important obstructions for synchronization despite the attractivity and the invariance of the synchronization manifold concern the possible limit sets of solutions which converge to the synchronization manifold and the asymptotic phase property of the synchronous solutions. Similar technical problems also appear in the study of asymptotically autonomous differential equations, see e.g. Strauss and Yorke [137]. Another area where similar issues play a major role is the study of the dynamics near manifolds of equilibria, see e.g. Aulbach [12].

In the following, we often utilize a special notation for a periodic solution associated with a periodic orbit Γ, more precisely:

Remark 2.35. *Consider an oscillator* (2.22) *with periodic orbit* Γ *for* $u = 0$. *By* $t \mapsto x(t; \gamma_0)$ *we denote a solution with an arbitrary but fixed initial condition* $\gamma_0 \in \Gamma$.

For our purposes it is important to check whether a periodic orbit has the asymptotic phase property. This can be done by studying the stability properties of the linearization of an oscillator about the solution $t \mapsto x(t; \gamma_0)$. More precisely if $\dot{x} = f(x, u)$ is the considered oscillator, $u = 0$ and $t \mapsto x(t; \gamma_0)$ is a periodic solution with period T, then the linearization about this solution is the differential equation

$$\dot{y} = \left. \frac{\partial f}{\partial x} \right|_{x=x(t;\gamma_0), u=0} y. \tag{2.108}$$

The system (2.108) is linear and time-varying with periodic coefficients. Let $t \mapsto \Phi(t)$ denote a fundamental matrix solution of (2.109), i.e. a matrix function whose columns form a fundamental solution, see e.g. Chicone [28, Definition 2.12]. Then Floquet's Theorem, see e.g. Chicone [28, Theorem 2.83], implies that the solution $t \mapsto y(t; y_0)$ of (2.108) satisfies

$$y(t; y_0) = P(t) \exp(tR) y_0 \tag{2.109}$$

where R is a matrix such that $\exp(TR) = \Phi^{-1}(0)\Phi(T)$ and where $P : \mathbb{R} \to \mathbb{R}^{n \times n}$ given by $\Phi(t) = P(t) \exp(tR)$ is periodic with period T, i.e. the same period as $t \mapsto x(t; \gamma_0)$. Alternative sources for Floquet's Theorem are e.g. Coddington and Levinson [31, Chapter 3.5] or Brockett [21, Chapter 1.8]. The eigenvalues of $\exp(TR)$, where T is the period of $x(t;\gamma_0)$, are called the *characteristic multipliers* of the system. If ρ is a characteristic multiplier, then $\mu \in \mathbb{R}$ is called a *characteristic exponent* if $\exp(\mu T) = \rho$, see Chicone [28, p. 192]. A well-known result is the following Theorem by Andronov and Witt:

Theorem 2.36 (Coddington and Levinson [31, Chap. 13, Theorem 2.2]). *Let $n - 1$ characteristic exponents of* (2.109) *have negative real parts, then the periodic solution $t \mapsto x(t;\gamma_0)$ is asymptotically orbitally stable and the associated periodic orbit Γ has the asymptotic phase property.*

Other references for this theorem are e.g. Hartman [56, Chap, IX, Theorem 11.1], Hale [53, Chap. IV, Theorem 2.1] or Chicone [28, Theorem 2.132]. Since the calculation of the characteristic exponents is computationally not always easy, we use a result of Muldowney [100], which transforms the problem of verifying the orbital stability for a periodic solution to a stability problem for a linear time-varying differential equation. For this we need the following definition.

Definition 2.37 (Compound matrices, Muldowney [100, I. Introduction]). *Let $A \in \mathbb{R}^{n \times n}$, and $A_{i_1,i_2}^{j_1,j_2}$ be defined by*

$$A_{i_1,i_2}^{j_1,j_2} = \det \begin{pmatrix} A_{i_1,j_1} & A_{i_1,j_2} \\ A_{i_2,j_1} & A_{i_2,j_2} \end{pmatrix}. \tag{2.110}$$

The second multiplicative compound $A^{(2)}$ of A is the $\binom{n}{2} \times \binom{n}{2}$ matrix whose entries are the $A_{i_1,i_2}^{j_1,j_2}$ ordered lexicographically, i.e. i_1,j_1 comes before i_2,j_2 if and only if $i_1 < i_2$ or $i_1 = i_2$ and $j_1 < j_2$. The second additive compound matrix $A^{[2]}$ of A is defined by

$$A^{[2]} = \frac{\partial}{\partial h}\bigg|_{h=0} (I + hA)^{(2)}. \tag{2.111}$$

The $A^{[2]}$ have a spectral property depending on the spectral properties of A which is of importance in the following.

Remark 2.38. *The eigenvalues of $A^{[2]}$ are the sums of all pairs of eigenvalues of A, see e.g. Muldowney [100, Section 2].*

For our investigations on the stability of (2.109) we use the following theorem.

Theorem 2.39 (Muldowney [100, Theorem 4.2]). *Let $t \mapsto x(t;\gamma_0)$ be a periodic solution of $\dot{x} = f(x)$ with period $T > 0$. If the linear system*

$$\dot{y} = \left(\frac{\partial f}{\partial x}\right)^{[2]}\bigg|_{x=x(t;\gamma_0)} y. \tag{2.112}$$

is asymptotically stable, then $x(t;\gamma_0)$ is asymptotically orbitally stable.

Remark 2.40. *The proof in Muldowney [100] shows that the asymptotic stability of* (2.112) *implies that the characteristic exponents are negative, i.e. the result also implies the asymptotic phase property.*

Remark 2.41. *The essential idea behind Theorem 2.39 is that the solution $t \mapsto y(t;y_0)$ of (2.112) is again of the form (2.109), i.e. $y(t) = P(t)^{[2]} \exp(tR^{[2]})$, where R is the matrix from (2.109). Since (2.112) is by assumption asymptotically stable, all eigenvalues of $R^{[2]}$ are negative. We assumed that $t \mapsto x(t;\gamma_0)$ is a periodic solution, hence there is a direction of neutral stability, i.e. there is one eigenvalue of R which is identical to zero. We use the fact that the eigenvalues of $R^{[2]}$ are sums of pairs of all different eigenvalues of R. Since one eigenvalue of R is identical to zero, all eigenvalues of R are also eigenvalues of $R^{[2]}$, i.e. beside the zero eigenvalue, R has only eigenvalues with negative real part. This implies that RT has beside one zero eigenvalue only eigenvalues with negative real parts. The eigenvalues of RT are the characteristic exponents of (2.109) and thus the Andronov-Hopf-Theorem 2.36 implies the asymptotic orbital stability of the periodic solution $t \mapsto x(t;\gamma_0)$ of $\dot{x} = f(x)$ and the asymptotic phase property for this solution and thus for the associated periodic orbit.*

2.3.3 Sufficient conditions for oscillator synchronization

In this section, we want to discuss the technical difficulties to infer synchronization in the sense of Definition 2.13. Furthermore, we give conditions which help us to infer synchronization for the networks of Lyapunov oscillators in Section 2.3.3.

We explain the technical difficulties in an abstract two-oscillator setup. Consider two oscillators of the form

$$\dot{x}_k = f(x_k, u_k) \tag{2.113}$$

with states $x_k \in \mathbb{R}^n$, inputs $u_k \in \mathbb{R}^p$ and $k \in \{1,2\}$. Both oscillators have identical vector fields and for $u_k = 0$ an asymptotically stable periodic orbit Γ with the asymptotic phase property. We assume that the periodic orbit Γ is the only stable limit set of (2.113) and that all unstable limit sets are isolated fixed points. Consider now two coupled oscillators where the coupling is given by a function $c(x_1,x_2)$ and where we set $u_1 = c(x_1,x_2)$ and $u_2 = -c(x_1,x_2)$, i.e. the equations for the oscillator network are

$$\begin{aligned} \dot{x}_1 &= f\big(x_1, c(x_1,x_2)\big) \\ \dot{x}_2 &= f\big(x_2, -c(x_1,x_2)\big). \end{aligned} \tag{2.114}$$

The condition $u_2 = -u_1$ is motivated from the class of so called *diffusive couplings*, which are often considered in literature, see e.g. Hale [54]. We assume that $x_1 = x_2$ implies $c(x_1,x_2) = 0$. Furthermore we assume $c(x_1,x_2)$ renders the set $\mathscr{S}$ defined as

$$\mathscr{S} = \{x \in \mathbb{R}^{2n} | x_1 = x_2\} \tag{2.115}$$

attractive and invariant. We aim for conditions that guarantee that the oscillators synchronize in the sense of Definition 2.13, i.e. that all solutions converge for all $x_0 \in D$ to a solution $t \mapsto \big(x_1(t+\delta;\gamma_0), x_1(t+\delta;\gamma_0)\big)$ for a suitable $\delta \in \mathbb{R}$ where $t \mapsto x_1(t;\gamma_0)$ is a periodic solutions of both uncoupled oscillators. Since we assumed that the coupling renders $\mathscr{S}$ attractive and invariant, these solutions fulfill $\lim_{t\to\infty} x_1(t;x_{01}) - x_2(t;x_{01}) = 0$. With the given assumptions we know that the asymptotic residual dynamics of the coupled oscillators with initial conditions in D are given by

$$\begin{aligned} \dot{x}_1 &= f(x_1,0) \\ \dot{x}_2 &= f(x_2,0). \end{aligned} \tag{2.116}$$

In other words, the residual dynamics are the individual oscillator dynamics. The question is now what we can infer from the knowledge of the dynamics of the single oscillators about the dynamics of the network of oscillators. By assumption we know the limit set of a point x_0 in the case of a single oscillator. Then $\omega(x_0)$ is either an asymptotically stable periodic orbit or one point in the finite set of isolated fixed points. The network of oscillators has twice the dimension of the original oscillator and the coupling $c(x_1, x_2)$ fulfills

$$x_1 = x_2 \Rightarrow c(x_1, x_2) = 0. \tag{2.117}$$

Hence, initial conditions of the form (x_0, x_0) with $x_0 \in \mathbb{R}^n$ imply that $x_1(t; x_0) \equiv x_2(t; x_0)$, where $t \mapsto x_k(t; x_0)$ is a solution of the uncoupled oscillator (2.113) with $u_k = 0$ for $k \in \{1, 2\}$. In other words the ω-limit set $\Omega\big((x_0, x_0)\big)$ is $(\Gamma \times \Gamma) \cap \mathscr{S}$ with $\mathscr{S}$ defined in (2.115), unless x_0 is a fixed point. Therefore, we know the dynamics of the oscillator network if the initial conditions lie in $\mathscr{S}$ and these dynamics have the desired synchronizing behavior for almost all initial conditions. The question is which properties guarantee the same for the remaining initial conditions. An answer to this question is the following Lemma, which is an immediate result.

Lemma 2.42. *Consider an oscillator network of the form (2.34) with state space $\mathbb{R}^{nN}$. Assume that the couplings c_k render the synchronization manifold $\mathscr{S}$ defined by*

$$\mathscr{S} = \big\{ (x_1, \ldots, x_N)^\top \in \mathbb{R}^{nN} | x_i = x_j \text{ for all } i, j \in \{1, \ldots, N\} \big\} \tag{2.118}$$

attractive and invariant. Then we achieve synchronization in the sense of Definition 2.13 if and only if there is a $D \subset \mathbb{R}^{nN}$ such that the following two conditions hold:

a) *For all initial conditions in $x_0 \in D$ the solutions $t \mapsto x(t; x_0)$ converge to the limit set $(\Gamma \times \ldots \times \Gamma) \cap \mathscr{S}$ (ω-limit set property for synchronization).*

b) *All solutions $t \mapsto x(t; x_0)$ which converge to $(\Gamma \times \ldots \times \Gamma) \cap \mathscr{S}$ have the property that the $t \mapsto x_k(t; x_0)$ converge to a periodic solution of the uncoupled oscillator (asymptotic phase property for synchronization).*

Proof. "$\Rightarrow$". Assume that we achieve synchronization in the sense of Definition 2.13, then for all $x_0 \in D$ we have $x_k(t; x_0) \to x_k(t; \gamma)$ for $t \to \infty$ where $\gamma \in \Gamma$ and Γ is a periodic orbit of the uncoupled oscillators. Hence for all $x_0 \in D$ the ω-limit set $\omega(x_0)$ is $(\Gamma \times \ldots \times \Gamma) \cap \mathscr{S}$, hence a) holds. Since $x_k(t; x_0) \to x_k(t; \gamma)$ for $t \to \infty$, the condition b) holds as well. "$\Leftarrow$". If the condition a) holds, then the ω-limit set of all $x_0 \in D$ is $(\Gamma \times \ldots \times \Gamma) \cap \mathscr{S}$. Because of b), for every x_0 such that $\omega(x_0) = (\Gamma \times \ldots \times \Gamma) \cap \mathscr{S}$ there is a $\gamma_{x_0} \in \Gamma$ such that $x_k(t; x_0) \to x(t; \gamma_{x_0})$ for $t \to \infty$, where $x(t; \gamma_{x_0})$ is a periodic solution for the uncoupled oscillator as explained in Remark 2.35. Since all solutions converge to $\mathscr{S}$ this γ_{x_0} is the same for all $k \in \{1, \ldots, N\}$. Hence for all $x_0 \in D$ and all $k \in \{1, \ldots, N\}$ we have $x_k(t; x_0) \to x(t; \gamma_{x_0})$ for $t \to \infty$. Therefore we have synchronization in the sense of Definition 2.13. ∎

Before discussing a class of oscillators where we can guarantee the ω-limit set property for synchronization (Lemma 2.42 a)) and the asymptotic phase property for synchronization (Lemma 2.42 b)), we discuss why these properties do not hold trivially if we assume that the coupling renders $\mathscr{S}$ attractive and invariant. First we discuss the ω-limit set property and then the asymptotic phase property of the synchronous solution. For illustration of problems concerning the ω-limit set property for synchronization (Lemma 2.42 a)) consider the illustration in Fig. 2.6 for two coupled oscillators. Although we know, as reasoned in the previous paragraph, that the ω-limit set $(\Gamma \times \Gamma) \cap \mathscr{S}$ is a limit set for all initial conditions within $\mathscr{S}$

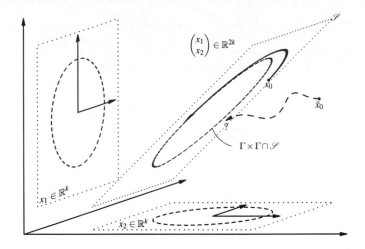

Figure 2.6: A sketch of the state space of the two coupled oscillators. The plane on the left illustrates the state space of $\dot{x}_1 = f(x_1,0)$ and the plane on the bottom the state space of $\dot{x}_2 = f(x_2,0)$. The plane in the middle illustrates $\mathscr{S}$, i.e. the state space of the coupled oscillators where $x_1 = x_2$. We know that solutions that start within $\mathscr{S}$ stay within $\mathscr{S}$ and that all non-fixed point solutions converge to the limit set of the synchronized solution, i.e. $(\Gamma \times \Gamma) \cap \mathscr{S}$. However, it is not guaranteed that a solution $t \mapsto \tilde{x}(t;\tilde{x}_0)$ starting from $\tilde{x}_0 \in D$ but $\tilde{x}_0 \notin \mathscr{S}$ converges to $(\Gamma \times \Gamma) \cap \mathscr{S}$.

which are not fixed points, we can not immediately guarantee that it is the only limit set for all initial conditions that converge to $\mathscr{S}$.

An example for the problems concerning the ω-limit set property is given in the following.

Example 2.43. *Consider the system*

$$\dot{r} = r(1 - r^2) + er^4$$
$$\dot{\phi} = 1 \tag{2.119}$$
$$\dot{e} = -e^3,$$

where the states r, ϕ, e fulfill $r \in \mathbb{R}_{>0}$, $\phi \in [0, 2\pi)$ and $e \in \mathbb{R}$, i.e. the system is given in polar coordinates. We consider the solution $t \mapsto (r(t;(r_0,\phi_0,e_0)), \phi(t;(r_0,\phi_0,e_0)), e(t;(r_0,\phi_0,e_0)))$ of (2.119) for two different cases, namely $e_0 = 0$ and $e_0 \neq 0$.

If the initial condition e_0 for e fulfills $e_0 = 0$ and $r_0 \in \mathbb{R}_{>0}$, $\phi \in [0, 2\pi)$, then the e-component of the solution has the property $e(t;(r_0,\phi_0,0)) = 0$ for all $t \geq 0$. In this case, the dynamics of r, ϕ are determined by

$$\dot{r} = r(1 - r^2)$$
$$\dot{\phi} = 1. \tag{2.120}$$

Therefore, all initial conditions of the form $(r_0,\phi_0,0)^\mathsf{T}$ converge to a circle with radius 1 and $t \mapsto \phi(t;(r_0,\phi_0,0)^\mathsf{T}) = t + \phi_0$, for more details see e.g. Minorsky [98]. In other words, solutions of (2.119) with initial conditions of the form $(r_0,\phi_0,0)$ have a nonempty ω-limit set.

However, for a given initial condition $e_0 \neq 0$, we always find an initial condition r_0 such that $r(t;(r_0,\phi_0,e_0)^{\mathsf{T}}) \to \infty$ in finite time although $e(t;(r_0,\phi_0,e_0)^{\mathsf{T}}) \to 0$ for $t \to \infty$ for all $e_0 \in \mathbb{R}$. Thus, the ω-limit set with such initial conditions is empty.

If we interpret Example 2.43 in the context of synchronization, then (2.119) represents the dynamics of the single oscillator and the e-dynamics would play the role of the vanishing coupling term due to $e = x_1 - x_2$. In this case the system would not have the ω-limit set property for synchronization (Lemma 2.42 a)). Even though the example is very simplified, it shows that the limit sets are not the same.

In the case of the asymptotic phase property for synchronization (Lemma 2.42 b)), the possible problems concern the dynamics of the solutions that reach the set $(\Gamma \times \Gamma) \cap \mathscr{S}$. More specifically, assume that Lemma 2.42 a) holds, i.e. all solutions with initial conditions in D also converge to $(\Gamma \times \Gamma) \cap \mathscr{S}$. Then, this does not imply that such a solution also converges towards a solution which has its initial condition on the limit set, i.e. a solution of the form $t \mapsto \big(x(t+c;\gamma_0), x(t+c;\gamma_0)\big)$ with $\gamma_0 \in \Gamma$. To illustrate the problem consider the situation depicted in Figure 2.7. An example for the problems concerning the asymptotic phase property

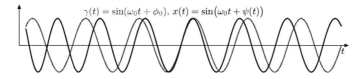

Figure 2.7: An illustration for the problems which prevent the synchronous solution to have the asymptotic phase property (Lemma 2.42 b)). Assume the solution $t \mapsto x(t;\gamma_0)$ of the uncoupled oscillator is a sinusoidal function, i.e. the periodic orbit of the uncoupled oscillator is a circle. Now assume that the asymptotic solution $t \mapsto x(t)$ of the coupled oscillator is given by $x(t) = \sin(\omega_0 t + \psi(t))$. Although both functions have the same limit set, i.e. a circle, it is still possible that the argument of the periodic function describing $t \mapsto x(t)$ has a function $\psi(t)$ which does not converge to a constant value for $t \to \infty$.

is given in the following.

Example 2.44. *Consider*

$$\dot{r} = r(1 - r^2)$$
$$\dot{\phi} = 1 + e^2 \tag{2.121}$$
$$\dot{e} = -e^3,$$

where the states r, ϕ, e fulfill $r \in \mathbb{R}_{>0}$, $\phi \in [0,2\pi)$ and $e \in \mathbb{R}$, i.e. the system is given in polar coordinates. We consider the solution $t \mapsto (r(t;(r_0,\phi_0,e_0)), \phi(t;(r_0,\phi_0,e_0)), e(t;(r_0,\phi_0,e_0)))$ of (2.119) for two different cases, namely $e_0 = 0$ and $e_0 \neq 0$.

If the initial condition e_0 for e fulfills $e_0 = 0$, we get $e(t;(r_0,\phi_0,0)) = 0$ for all $t \geq 0$ and the dynamics of r, ϕ are determined by (2.120). This implies that for $e_0 = 0$ the solution $t \mapsto (r(t;(r_0,\phi_0,0)), \phi(t;(r_0,\phi_0,0)), e(t;(r_0,\phi_0,0)))$ converges to a unit circle S^1 and moves on S^1 with unit speed.

If we choose a nonzero initial condition $\tilde{e}_0 \neq 0$, then we observe a different behavior. First, note that $e(t;(r_0,\phi_0,\tilde{e}_0)) = |\tilde{e}_0| (2\tilde{e}_0^2 t + 1)^{-\frac{1}{2}}$. Hence for any ϕ_0 we get $\phi(t;(r_0,\phi_0,\tilde{e}_0)) = \psi(t) + t \pmod{2\pi}$ with $\psi(t) = \frac{1}{2}\ln(2t\tilde{e}_0^2 + 1) + \phi_0$. As a consequence, we know $\phi(t;(r_0,\phi_0,\tilde{e}_0)) -$

$\phi\big(t;(r_0,\phi_0,0)\big) = \psi(t)$ (mod 2π). *Since the r-dynamics did not change we still converge to the circle with radius μ, but since ψ does not converge to a fixed value for $t \to \infty$, there is no constant c such that*

$$\lim_{t\to\infty} \phi(t;(r_0,\phi_0,\tilde{e}_0)) - \phi(t;(r_0,\phi_0,0)) = c. \qquad (2.122)$$

Hence, no periodic solution $t \mapsto (r(t;(1,\phi_0,0)),\phi(t;(1,\phi_0,0)),e(t;(1,\phi_0,0)))$ does have the asymptotic phase property from Definition 2.3.

If we interpret Example 2.44 in the context of synchronization, then (2.121) would play the role of a single oscillator and the e-dynamics would model the vanishing coupling term due to $e = x_1 - x_2$. In this case the asymptotic phase property for synchronization (Lemma 2.42 b)) would be violated.

The two examples illustrated that a vanishing error in the state differences of the oscillators is only necessary to achieve synchronization but not sufficient: a vanishing error does neither guarantee the ω-limit set property for synchronization (Lemma 2.42 a)) nor the asymptotic phase property for synchronization (Lemma 2.42 b)).

Above we showed that convergence towards the synchronization manifold does in general not imply synchronization. In the following, we show that a network of Lyapunov oscillators is a system class where we can guarantee the ω-limit set property for synchronization (Lemma 2.42 a)). Furthermore we give an additional property that guarantees the asymptotic phase property for synchronization (Lemma 2.42 b)). These results are given in the Lemmata 2.45 and 2.47, and the Theorem 2.49. Lemma 2.45 shows that the stability properties of the individual Lyapunov oscillators together with the coupling that vanishes on $\mathscr{S}$ implies the ω-limit set property for synchronization (Lemma 2.42 a)).

Lemma 2.45. *Given a network of N Lyapunov oscillators (Definition 2.32) with coupling functions $c_k : \mathbb{R}^{nN} \to \mathbb{R}^p$, i.e.*

$$\dot{x}_k = f\big(x_k, c_k(x_1,\ldots,x_N)\big). \qquad (2.123)$$

Denote the Lyapunov function associated with oscillator k by $W_k : \mathbb{R}^n \to \mathbb{R}$. Let the synchronization manifold $\mathscr{S}$ be defined by (2.118). Assume that $c_k(x) = 0$ for $x \in \mathscr{S}$ and $k \in \{1,\ldots,N\}$. Furthermore assume that for $x_0 \in \mathbb{R}^{nN}$ the solution $t \mapsto x(t;x_0)$ of (2.123) is bounded and that $x(t;x_0) \to \mathscr{S}$ for $t \to \infty$. Then the ω-limit set $\omega(x_0)$ for all initial conditions $x_0 \in \mathbb{R}^{nN} \setminus (F \times \ldots \times F)$ is given by $(\Gamma \times \ldots \times \Gamma) \cap \mathscr{S}$.

Proof. We want to show for a given initial condition $x_0 \in \mathbb{R}^{nN}$ with bounded solution $t \mapsto x(t;x_0)$ and with $x(t;x_0) \in \mathscr{S}$ for $t \to \infty$, that the ω-limit set fulfills $\omega(x_0) = (\Gamma \times \ldots \times \Gamma) \cap \mathscr{S}$. Our proof utilizes the general result on the asymptotic behavior of solutions of differential equations which converge to an invariant manifold, see Section 1.2.2. We construct a height function W on $\mathscr{S}$ and consider the (topologically) connected components $\{\mathscr{E}_i\}_{i\in I}$ of the set $\mathscr{E}$, defined as

$$\mathscr{E} = \{x \in \mathscr{S} | \dot{W}(x) = 0\}. \qquad (2.124)$$

Then we show that all $\mathscr{E}_i$ lie within a level set of W which allows us to conclude by Arsie and Ebenbauer [9, Theorem 6] that $\omega(x_0) \subset \mathscr{E}_i$ for a unique $i \in I$.

We know that the solutions we consider are bounded. Therefore, $\omega(x_0)$ is nonempty, connected, compact and invariant and is the smallest set approached by the solution $t \mapsto x(t;x_0)$, see LaSalle [80, Chapter 2, Theorem 5.2].

Consider $W : \mathbb{R}^{nN} \to \mathbb{R}$ defined by

$$W(x) = \sum_{k=1}^{n} W_k(x_k). \tag{2.125}$$

W is defined on $\mathbb{R}^{nN}$ and is C^1. Since W is a sum of non-negative functions, W is non-negative. More specific, $W_k(x_k)$ is zero if and only if $x_k \in \Gamma$, hence $W(x) = 0$ if and only if $x \in \Gamma \times \ldots \times \Gamma$. The derivative of W in direction of the vector field (2.123) is given by

$$\dot{W}(x) = \sum_{i=1}^{n} \dot{W}_k(x_k) = \sum_{i=1}^{n} \frac{\partial W}{\partial x_k}(x_k) f(x_k, c_k(x_1, \ldots, x_N)). \tag{2.126}$$

$\dot{W}$ is not definite for $x \in \mathbb{R}^{nN}$, only for $x \in \mathscr{S}$. More precisely, for $x \in \mathscr{S}$ the coupling functions fulfill $c_k(x) = 0$. Furthermore, if $x \in \mathscr{S}$, then $\dot{W}$ is the sum of the derivatives of W_k in direction of the vector fields of the respective uncoupled oscillator at x_k, i.e. according to Definition 2.32.c)ii) the sum of $\dot{W}_k(x_k) = \frac{\partial W_k}{\partial x_k}(x_k) f(x_k, 0)$. Since we consider Lyapunov oscillators, the property Definition 2.32.c)i) implies that $\dot{W}_k(x_k) = 0$ either if $x_k \in \Gamma$ or if $x_k \in F$, where F denotes a finite set of isolated fixed points. Furthermore, we know that $W_k(x_k) > 0$ for all $x_k \in F$ and $\dot{W}_k(x_k) = 0$, $(x_k \in F)$. Utilizing $\mathscr{E}$ as defined in (2.124), we thus get that $W : \mathbb{R}^{nN} \to \mathbb{R}$ fulfills $\dot{W}|_{\mathscr{S}} \leq 0$ and $\dot{W}|_{\mathscr{S} \backslash \mathscr{E}} < 0$. In other words, W is a *height function* for the pair $(\mathscr{S}, [f, \ldots, f])$.

Now consider the connected components $\{\mathscr{E}_k\}_{k \in I}$ of $\mathscr{E}$, where I is a suitable index set. To apply the result from Arsie and Ebenbauer [9], we have to show that each $\mathscr{E}_k$ lies in a level set of W and $\{W(\mathscr{E}_k)\}_{k \in I} \subset \mathbb{R}$ has at most a finite number of accumulation points.

Since $\mathscr{E} = \{x \in \mathscr{S} | \dot{W}(x) = 0\}$ and $\mathscr{S} = \{x \in \mathbb{R}^{nN} | x_i = x_j, \ i, j \in \{1, \ldots, N\}\}$, we know that $\mathscr{E} = ((\Gamma \times \ldots \times \Gamma) \cup (F \times \ldots \times F)) \cap \mathscr{S}$. Since W is the sum of the W_k and $W_k|_{\Gamma} = 0$, all $\mathscr{E}_k$ lie within level sets of W. Furthermore, $\{W(\mathscr{E}_k)\}$ is a finite set, thus it fulfills the assumptions necessary to apply Theorem 1.17 (Arsie and Ebenbauer [9, Theorem 6]). As a consequence, $\omega(x_0) \subset \mathscr{E}_k$ for a unique $k \in I$. Because of our assumptions, $(\Gamma \times \ldots \times \Gamma) \cap \mathscr{S}$ is the only asymptotically stable set within $\mathscr{E}$. The other points in $\mathscr{E}$ are exponentially anti-stable. Therefore, an initial condition starting in $\mathbb{R}^{nN} \backslash (F \times \ldots \times F)$ converges towards $(\Gamma \times \ldots \times \Gamma) \cap \mathscr{S}$. ∎

Remark 2.46. *It is possible to proof Lemma 2.45 in a local setting, i.e. if the periodic orbit of the oscillators is not globally stable, but has a region of attraction $\tilde{D}$. Then the additional requirement is that $x(t; x_0) \to (\tilde{D} \times \ldots \times \tilde{D}) \cap \mathscr{S}$, $(x_0 \in \tilde{D})$.*

Lemma 2.45 therefore provides conditions for the coupling functions c_i that guarantee the ω-limit set property (Lemma 2.42 a)) for a network of Lyapunov oscillators. In the next lemma we consider conditions for the coupling that allow us to conclude that the asymptotic phase property (Lemma 2.42 b)) holds.

Lemma 2.47. *Given a network of N Lyapunov oscillators with coupling $c_i : \mathbb{R}^{nN} \to \mathbb{R}^p$, i.e. a system of the form (2.123). Denote the vector field of the coupled system by g, i.e. $g : \mathbb{R}^{nN} \to \mathbb{R}^{nN}$ with $g_k(x_1, \ldots, x_n) = f(x_k, c_k(x_1, \ldots, x_n))$ for $k \in \{1, \ldots, N\}$. If the linear time-varying system*

$$\dot{y} = \left(\frac{\partial g}{\partial x}\right)^{[2]} \Bigg|_{(x_1, \ldots, x_n) = (x(t; \gamma_0), \ldots, x(t; \gamma_0))} y \tag{2.127}$$

is asymptotically stable, then property Lemma 2.42 b) holds for (2.123).

Proof. By Theorem 2.39 the asymptotic stability of (2.127) is sufficient for the asymptotic phase property of the solution $t \mapsto \left(x(t;\gamma_0),\ldots,x(t;\gamma_0)\right)$. We know that the synchronization manifold $\mathscr{S}$ is positively invariant and by Lemma 2.45 all solutions with initial conditions in $D \setminus (F \times \ldots \times F)$ converge towards $(\Gamma \times \ldots \times \Gamma) \cap \mathscr{S}$. Thus, for a solution $t \mapsto x(t;x_0)$ which converges to $\mathscr{S} \cap (\Gamma \times \ldots \times \Gamma)$ for $t \to \infty$, there is an $M > 0$ and a $T > 0$ such that $\left| s(t;x_0) - \left(x(t;\gamma_0),\ldots,x(t;\gamma_0)\right) \right| < M$ for all $t \geq T$ and then we know by Coddington and Levinson [31, p.323, Theorem 2.2] that there is a $c \in \mathbb{R}$ such that

$$\lim_{t \to \infty} \left| s(t;x_0) - \left(x(t+c;\gamma_0),\ldots,x(t+c;\gamma_0)\right) \right| = 0. \tag{2.128}$$

∎

Remark 2.48. *The linear system (2.127) is time varying, hence it is not straight-forward to check asymptotic stability of (2.127). There are stability results which involve norm bounds and bounds on the eigenvalues of the system matrix which are easy to compute, but most of these results are conservative. However it is possible to check the asymptotic stability of (2.127) in a practical way with the help of a numerical integration of (2.127). A solution of (2.127) has the form $y(t;y_0) = P(t)\exp(tR)y(0)$ where $P : \mathbb{R} \to \mathbb{R}^{n \times n}$ is T-periodic and T is the period of the solution $t \mapsto x(t;\gamma_0)$. In other words, the transition matrix is $\Phi(t,t_0) = P(t)\exp(tR)$. We are able to determine the transition matrix numerically, since we can determine $Y(t) \in \mathbb{R}^{\binom{n}{k} \times \binom{n}{k}}$ for $Y(t_0) \in \mathbb{R}^{\binom{n}{k} \times \binom{n}{k}}$ by $Y(t) = \Phi(t,t_0)Y(t_0)$. Since we know the period T and that $\Phi(T,0) = P(T+0)\exp(TR) = P(0)\exp(TR)$ and $\Phi(0,0) = id = P(0)\exp(0R) = P(0)$, we have $\Phi(T,0) = \exp(TR)$. Therefore, we can determine the eigenvalues of R and thus the stability of (2.127).*

With the preparation of Lemma 2.45 and Lemma 2.47, we are now able to formulate our final result for synchronization concerning networks of Lyapunov oscillators.

Theorem 2.49. *Consider a network of N Lyapunov oscillators with coupling functions c_k : $\mathbb{R}^{nN} \to \mathbb{R}^p$ for $k \in \{1,\ldots,N\}$, i.e. (2.123). Assume that $c_k(x) = 0$ for $x \in \mathscr{S}$ and $k \in \{1,\ldots,N\}$. Furthermore assume that for $x_0 \in \mathbb{R}^{nN}$ the solution $t \mapsto x(t;x_0)$ of (2.123) is bounded and that $x(t;x_0) \to \mathscr{S}$ for $t \to \infty$. Moreover suppose that (2.127) is asymptotically stable, then (2.123) synchronizes in the sense of Definition 2.13.*

Proof. By assumption we know that property Lemma 2.42 a) holds. Lemma 2.45 guarantees that the limit set is $(\Gamma \times \ldots \times \Gamma) \cap \mathscr{S}$, i.e. that the solution of (2.123) converges towards a solution $t \mapsto \left(x\left(\left(\psi(t);\gamma_0\right),\ldots,x\left(\psi(t);x_0\right)\right)\right)$, where $t \mapsto x(t;\gamma_0)$ is the periodic solution of the uncoupled Lyapunov oscillators and ψ is a function. Lemma 2.47 finally guarantees that there exists a $c \in \mathbb{R}$ such that $\psi(t) = t + c$. Therefore property Lemma 2.42 b) holds. ∎

Remark 2.50. *According to Remark 2.46, we are also able to state the result of Theorem 2.49 in a local setting, i.e. if the periodic orbits of the Lyapunov oscillators have a region of attraction $\tilde{D}$ which is not $\mathbb{R}^n$.*

In the following section we are going to consider an example that illustrates our results.

2.3.4 Example for the synchronization of two planar Lyapunov oscillators

We consider the synchronization of two systems of the form

$$\dot{x}_k = -y_k + x_k\left(\mu^2 - (x_k^2 + y_k^2)\right)$$
$$\dot{y}_k = x_k + y_k\left(\mu^2 - (x_k^2 + y_k^2)\right) \tag{2.129}$$

with states $x_k, y_k \in \mathbb{R}$ for $k \in \{1,2\}$. We denote in the following the vector field of (2.129) by g. The function $t \mapsto (x_k(t;\gamma_0); y_k(t;\gamma_0))^\mathsf{T}$ with $x_k(t;\gamma_0) = \mu\cos(t)$ and $y_k(t;\gamma_0) = \mu\sin(t)$ is a solution to (2.129) with the initial condition $\gamma_0 = (\mu,0)$. The ω-limit set of $\gamma_0 = (\mu,0)$ is the circle with radius μ in the following abbreviated by S_μ^1.

Consider $W : \mathbb{R}^2 \to \mathbb{R}, (x_k, y_k) \mapsto \frac{1}{4}(x_k^2 + y_k^2 - \mu)^2$. W is strictly positive for all $(x_k, y_k) \notin S_\mu^1$. Furthermore the derivative $\dot{W}$ of W along the trajectories of (2.129) evaluate to

$$\dot{W}(x,y) = (x_k^2 + y_k^2 - \mu)(x_k\dot{x}_k + y_k\dot{y}_k)$$
$$= -(x_k^2 + y_k^2)(\mu - (x_k^2 + y_k^2))^2 \leq 0.$$

The derivative of W along the trajectories of (2.129) is zero for $x_k^2 + y_k^2 = \mu$ and $(x_k, y_k) = (0,0)$ and no other points. Therefore, according to the theory of stability of invariant set Zubov [152, Theorem 12], S_μ^1 is asymptotically stable, and the solutions $s(t; x_0)$ of all initial conditions $x_0 \in \mathbb{R}^2 \setminus \{0\}$ converge towards S_μ^1. We check the orbital stability of $x(t; \gamma_0)$ with the help of Theorem 2.39. As a consequence, we calculate $(x_k(t), y_k(t)) = 2\mu^2 - 4(x_k^2 + y_k^2)$. Since $(\frac{\partial g}{\partial x})^{[2]}(x(t; \gamma_0)) = -2\mu^2$ and $\mu > 0$, we know that $\dot{z} = (\frac{\partial g}{\partial x})^{[2]}(x(t; \gamma_0))z$ is asymptotically stable, and we thus have orbital stability according to Theorem 2.39. Therefore system (2.129) is a Lyapunov oscillator in the sense of Definition 2.32.

Our goal is now to find a synchronizing coupling for two systems of the form (2.129), i.e. to find coupling functions $c_1(x_1, x_2, y_1, y_2)$, $c_2(x_1, x_2, y_1, y_2)$ such that

$$\dot{x}_1 = -y_1 + x_1\left(1 - (x_1^2 + y_1^2)\right) + c_1(x_1, x_2, y_1, y_2)$$
$$\dot{y}_1 = x_1 + y_1\left(1 - (x_1^2 + y_1^2)\right) + c_2(x_1, x_2, y_1, y_2)$$
$$\dot{x}_2 = -y_2 + x_2\left(1 - (x_2^2 + y_2^2)\right) - c_1(x_1, x_2, y_1, y_2)$$
$$\dot{y}_2 = x_2 + y_2\left(1 - (x_2^2 + y_2^2)\right) - c_2(x_1, x_2, y_1, y_2). \tag{2.130}$$

synchronizes for almost all initial conditions and that the synchronous solution has the form $t \mapsto (\cos(t + \phi_0), \sin(t + \phi_0), \cos(t + \phi_0), \sin(t + \phi_0))^\mathsf{T}$ for a $\phi_0 \in [0, 2\pi)$. Note that we restrict ourselves to $\mu = 1$ since we can transform the original equations by $\xi = \mu x$ and $\eta = \mu y$ to obtain the normalized equations (2.130).

We design the coupling by utilizing that the vector fields of (2.130) are polynomial. Therefore, we transform the equations (2.130) into the coordinates $e_1 = x_1 - x_2$, $e_2 = y_1 - y_2$, $s_1 = x_1 + x_2$ and $s_2 = y_1 + y_2$ and rewrite (2.130). Using $e = (e_1, e_2)$, $s = (s_1, s_2)^\mathsf{T}$ and $c = (c_1, c_2)^\mathsf{T}$, we obtain a system of the form

$$\dot{e} = h(e,s) + c(e,s)$$
$$\dot{s} = m(e,s). \tag{2.131}$$

Since we have polynomial system equations and a linear transformation, the error dynamics are also polynomial in e and s, i.e. $h(e,s)$ and $m(e,s)$ are polynomials in e and s. Our goal is to find $c(e,s)$ such that (2.131) is asymptotically stable with respect to e and such that $e = 0$ implies $c(e,s) = 0$. In Lyapunov terms this means that we search for a Lyapunov function V with $V(e) > 0$ and and controller c as a function of (e,s) such that

$$\dot{V}(e) = \frac{\partial V}{\partial e}\big(h(e,s) + c(e,s)\big) < 0, \ (s \in \mathbb{R}^2). \tag{2.132}$$

Note that (2.132) is bilinear in the unknowns $V(e)$ and $c(e,s)$ and therefore does not allow a formulation which admits an efficient solution scheme.

However, in the present problem a quadratic polynomial ansatz for V, i.e. $V(e) = \frac{1}{2}(e_1^2 + e_2^2)$, together with a polynomial ansatz with a degree from one to three in e and s for c leads to a solution of (2.132) with the help of the sum of squares method, see e.g. Parrilo [108]. More specifically, if our ansatz for V is $V(e) = e^{\mathsf{T}}Qe$ with Q diagonal, then the second dissipation inequality is $e^{\mathsf{T}}Qh(e,s) + e^{\mathsf{T}}Qc(e,s) < 0$. Substituting $l(e,s)$ for $Qc(e,s)$, we obtain $e^{\mathsf{T}}Qh(e,s) + e^{\mathsf{T}}l(e,s)$, which is linear in all unknowns. The result for $c(e,s)$ is

$$\begin{aligned}
c_1(e,s) = &- 1.595e_1 + .524e_2 + .573e_1^3 - 1.749e_1^2e_2 \\
&+ .412e_1e_2^2 - 2.647e_2^3 + 1.446e_1s_1^2 - 2.68e_2s_1^2 \\
&+ .275e_1s_1s_2 + .789e_2s_1s_2 + .551e_1s_2^2 - 2.60e_2s_2^2 \\
c_2(e,s) = &- 2.e_1 - .211e_2 - .89e_1^3 + .073e_1^2e_2 \\
&- .235e_1e_2^2 + .051e_2^3 - .566e_1s_1^2 + .185e_2s_1^2 \\
&- .012e_1s_1s_2 + .041e_2s_1s_2 - .86e_1s_2^2 + .07e_2s_2^2.
\end{aligned} \tag{2.133}$$

We apply stability analysis of the compound matrix with the numerical method described in Remark 2.48. This yields a spectrum with strictly negative real part for R, where R is the matrix in the solution (2.109) of the compound system (2.127), more precisely

$$\begin{aligned}
\sigma(R) = \{&-4.04 + 0.19\mathrm{j}, -4.04 - 0.19\mathrm{j}, -4.22 + 0.26\mathrm{j}, \\
&-4.22 - 0.26\mathrm{j}, -4.59, -4.41\},
\end{aligned} \tag{2.134}$$

where j denotes the imaginary unit. By Theorem 2.49, we thus know that all solution of (2.130) synchronize as long as they do not start in a fixed point.

2.3.5 Summary

We investigated synchronization in a network of Lyapunov oscillators. By synchronization we mean that a coupling between the oscillators drives all oscillators towards a common periodic solution. We showed that attractivity and invariance of the synchronization manifold is not sufficient for synchronization. We presented conditions that are sufficient for synchronization in oscillator networks. Then we showed that these sufficient conditions are fulfilled in a network of Lyapunov oscillators. More specifically, we showed in Lemma 2.45, that the assumption of a Lyapunov function for the periodic orbits of the individual oscillators allows us to conclude that all solutions which do not start in fixed points and which converge towards the synchronization manifold also converge towards the synchronous periodic orbit, i.e. the solutions have the ω-limit set property (Lemma 2.42 a)). Furthermore we gave in a condition for the coupling in Lemma 2.47, which implies that the solutions which converge to the synchronous periodic orbit have the asymptotic phase property (Lemma 2.42 b)). In literature, these properties are often assumed to hold a priori, although there is less known how to choose a coupling that guarantees these properties. The remaining open problem is to find synthesis methods for couplings for specific system classes that guarantee the ω-limit set property (Lemma 2.42 a)) and the asymptotic phase property (Lemma 2.42 b)). We sketched a possible way for such methods in the two-oscillator example.

2.4 Discussion

In this section, we considered synchronization of oscillators in oscillator networks from two perspectives.

In Section 2.2, we considered a specific class of phase model, the so called Kuramoto model, and investigated the influence of delays on frequency synchronization and phase synchronization. We derived simple necessary algebraic conditions for both synchronization phenomena. In the case of frequency synchronization of delayed Kuramoto models with an all-to-all graph associated to the delayed Kuramoto model, we could give explicit sufficient conditions for frequency synchronization depending on the system parameters. Our findings showed that larger coupling constants allow us to compensate the effect of the delays. The analysis provided for frequency synchronization in delayed Kuramoto models with an associated all-to-all graph allowed us to improve the existing sufficient conditions for frequency synchronization in undelayed Kuramoto models with associated all-to-all graphs. The conditions we obtained for the undelayed case are supported by the findings in Dörfler and Bullo [35]. In addition to that we could show that the existence of a solution to the necessary self-consistency conditions allows to give sufficient conditions for phase synchronization and frequency synchronization in delayed Kuramoto models in the case of an arbitrary associated directed graph which contains a directed spanning tree. We demonstrated that phase synchronization is possible in delayed Kuramoto models, even if the frequencies of the individual oscillators are non-identical and provided a numerical example for this. For undirected networks this was already observed in Papachristodoulou et al. [107]. Note that phase-synchronization is not possible in the undelayed case, which is one notable difference of the dynamics between the delayed and the undelayed case. The simulation example we provided in Section 2.2.4 shows that the obtained conditions in these cases are conservative. Furthermore, to apply our results for the arbitrary graph structure, one needs a solution to the self-consistency condition, which might be very hard to obtain. Our analysis results are not global, i.e. hold only for a subset of the possible initial conditions. It is desirable to remove this limitation and to provide a global analysis for the delayed Kuramoto model. In other words provide a full analysis of the model on the N-dimensional torus T^N. Note that for frequency synchronization in the undelayed case, this problem is also open, even if the associated graph is all-to-all.

In contrast to the results given for the Kuramoto model, the analysis given in Section 2.3 directly considers oscillator networks. We analyzed networks of identical oscillators where we assumed the existence of Lyapunov functions for the periodic orbits. Furthermore, we assumed that the couplings render the synchronization manifold $\mathscr{S}$ attractive and that the couplings are zero on $\mathscr{S}$. In other words we assume that $\mathscr{S}$ is attractive and invariant and discuss in detail why this is not sufficient for synchronization in oscillator networks. We identified two additional properties, i.e. the ω-limit set property for synchronization (Lemma 2.42 a)) and the asymptotic phase property for synchronization (Lemma 2.42 b)) which are necessary and sufficient for synchronization if the couplings renders $\mathscr{S}$ attractive and invariant. Finally, we utilize the Lyapunov functions for the periodic orbits of the individual oscillators to prove that the ω-limit set property for synchronization holds in a network of Lyapunov oscillators.

An additional test of the exponential orbital stability of the synchronous solution ensures the asymptotic phase property for synchronization. Our analysis shows that the presence of additional equilibrium points for the uncoupled oscillators dynamics does not change this result, as long as the other equilibria are anti-stable. It should be possible to relax these conditions further. More precisely, this should hold as long as almost all initial conditions in a uncoupled oscillator converge to the periodic solution. The provided result is an analysis result. It would be desirable to derive corresponding synthesis results, i.e. give methods to find coupling functions which ensure our properties that lead to synchronization. The problem to design a coupling that renders $\mathscr{S}$ attractive can be seen as the asymptotic stabilization of a set in a nonlinear system, which is in general a hard problem. This has to be addressed separately for specific classes of nonlinear systems. In our two-oscillator example we demonstrated that the class of polynomial systems could be an interesting candidate. Every proposed approach should also ensure the asymptotic phase property for the synchronous solution. Finding a method to synthesize a controller which guarantees the asymptotic phase property for periodic solutions in control system would also be very useful outside of the synchronization context.

3

Global output regulation and the regulation of rigid bodies

In the second part of the thesis we discuss a class of output regulation problems from a global perspective. Output regulation problems are control problems where the system is affected by a family of disturbances. The objective is that an output asymptotically tracks a family of references while asymptotically rejecting all disturbances from a given family. This is a fundamental problem in control theory, for additional background see e.g. Byrnes and Isidori [24] or Brockett [22]. If we consider the error between the output and the desired reference as new output, the so called *regulated output*, then the above problem is equivalent to achieve convergence of the regulated output to a constant value for all disturbances from the given family. One well established approach to solve such problems is based on the assumption that the references or disturbances acting on the control system are the solutions of a known differential equation, the so called *exosystem*, see e.g. Isidori and Marconi [67]. The classical solution of the output regulation problem for linear systems utilizing the internal model principle is due to Francis and Wonham, see Francis and Wonham [44]. For some classes of nonlinear systems, a well known solution for local output regulation problems is due to Isidori and Byrnes, see Isidori and Byrnes [65], Isidori [64]. Various extensions of this problem were considered in Serrani et al. [131; 132], Byrnes and Isidori [23], Huang [60], Pavlov et al. [109].

Here, we treat a specific class of output regulation problems where the non-trivial geometry of the state space poses a challenge for established design approaches. The considered problems are output regulation problems for systems on the special Euclidean group $SE(n) = \mathbb{R}^n \times SO(n)$ with $SO(n) = \{\Theta \in \mathbb{R}^{n \times n} | \Theta^{-1} = \Theta^T, \det \Theta = 1\}$ and for rigid body systems. Both problems are related since the configuration manifold for the rotational motion of a rigid body is $SO(3)$. The general importance of control problems where the non-trivial geometry of the state space is an obstruction for control design is well known, see e.g. Bhat and Bernstein [15] or Koditschek [73]. Especially for the rotational motion of a rigid body there is a large body of literature where control problems are considered, which take the non-trivial geometry into account. Due to the large number of different setups and different solution approaches, we only list here the recent contributions Sanyal et al. [116], Fernandes Vasconcelos et al. [42], Mayhew and Teel [91] and the overview Chaturvedi et al. [26], which contains some explanations on the topological problems for the rigid body case. For output regulation problems where the geometry of the state-space is non-trivial the existing solution approaches are local or semi-global, for problems concerning the rotational rigid body motion see e.g. Isidori et al. [68] or Chen and Huang [27].

In this thesis, we present a novel global solution approach for the considered class of output regulation problems on $SE(n)$ and for rigid body systems. In each case, we propose a feedback which guarantees the convergence to a finite number of isolated invariant sets which are unstable except one, which is asymptotically stable. Furthermore, we propose an observer with similar convergence properties. The feedback and the observer are designed independently. We show that the desired invariant set is the only asymptotically stable set in the closed loop with a *certainty equivalence implementation* of the feedback, i.e. the closed loop consists of the control system, the observer and the feedback utilizing the estimated states. As a consequence, we obtain a nonlinear separation principle for the considered problem classes. Preliminary versions of the results in this Chapter were published in Schmidt et al. [120; 122]. Parts of the results in Section 3.2 will be published in Schmidt et al. [123]. The results in Section 3.3 were published in Schmidt et al. [124]. The results in Section 3.4 were made publicly available in Schmidt et al. [121].

This chapter is structured as explained in the following. We explain one important challenge for the considered class of output regulation problems in Section 3.1 and explain how the presented solution approach resolves the challenge. In Section 3.2, we present a class of output regulation problem on the special Euclidean group $SE(n)$ and a global solution to the problem. In Section 3.3 we consider a class of output regulation problems for the rotational rigid body dynamics, present a global solution to the problem and consider an application scenario for a multibody satellite. We close the chapter in Section 3.4, where we discuss the obtained results.

3.1 Observer-based output regulation for systems with multiple equilibria

Output regulation problems have a long history in control theory, see e.g. Byrnes and Isidori [24] or Brockett [22]. In the general formulation of the nonlinear output regulation problem as discussed in Isidori [64], Isidori and Marconi [67], a control system is given by

$$\dot{x} = F(w,x,u)$$
$$e = H(w,x) \tag{3.1}$$

There, $x \in \mathbb{R}^n$ is the state, $e \in \mathbb{R}^q$ is the *regulated output*, $u \in \mathbb{R}^p$ is the controlled input and $w \in \mathbb{R}^m$ is the *exogenous* input of the system. The exogenous input includes disturbances, unknown references and unknown parameters.

An established assumption for many solution approaches to the output regulation is that the exogenous inputs w are solutions of a (neutrally stable) differential equation

$$\dot{w} = s(w). \tag{3.2}$$

The knowledge that the input signal $w : \mathbb{R} \to \mathbb{R}^m$ is the solution of a system (3.2) is not uncommon for design problems. This assumption is a trade-off between the advantageous but unrealistic situation that w is available as real-time measurement and the situation that nothing is known about w, for more details consult Isidori and Marconi [67].

The problem of nonlinear output regulation is to design a controller of the form

$$\dot{x}_c = F_c(x_c, e)$$
$$u = H_c(x_c, e), \tag{3.3}$$

where $x_c \in \mathbb{R}^\nu$, such that all solutions of the closed loop system (3.1), (3.2) and (3.3) remain bounded and such that $e(t) \to 0$ as $t \to 0$.

A necessary condition for the existence of a solution of the output regulation problem is that the state space of the closed loop consisting of (3.1), (3.3) and (3.2) contains a subset where the regulated output takes the desired value. If a controller renders a part of this subset invariant and achieves convergence to this part, then the output regulation problem is solved. In order to achieve the invariance of a subset where the regulated output takes the desired value, it is necessary that the controller contains a model of the exosystem. In other words, a controller that achieves asymptotic regulation must contain the dynamics of the exosystem, see Francis and Wonham [44] or Wonham [150, Chapter 8] for the linear case and Isidori and Byrnes [65] and Byrnes and Isidori [23] for the nonlinear case. This well known property is the so-called *internal model principle*.

In the case that the full state of the control system and the exosystem are available from measurements, a feedback depending on the states of the control system and the exosystem solves the output regulation problem. Since some states of the exosystem represent disturbances, it is common that at least these states are not available from measurements. Then, the question is how to incorporate the dynamics of the exosystem into the controller. One solution approach in that case is based on an observer which estimates the states of the control system and the exosystem, see e.g. Huang [60, Section 3.3]. We utilize this solution approach and design an output feedback controller that consists of an observer and the state feedback utilizing the estimated states, i.e. a certainty equivalence implementation of the state feedback, see e.g. Kokotovic [74] or Praly and Arcak [112]. The challenge for certainty equivalence controllers in the case of nonlinear control systems is the lack of a general separation principle as it is known for linear systems. More precisely, for linear systems it is well-known that a convergent observer together with a stabilizing state feedback which utilizes the estimated states guarantees closed loop stability, see e.g. Luenberger [86] or Kailath [72, Chapter 4.2]. For nonlinear control systems, two main obstructions for the certainty equivalence controller are finite escape time of the closed loop solutions and the presence of multiple isolated equilibria due to the geometry of the state space. The problem of finite escape time is discussed for example in Mazenc et al. [92]. Exemplary design approaches to deal with the problem of finite escape time utilize high-gain designs as in Atassi and Khalil [11], or rely on state feedback designs which achieve input-to-state stability with respect to the observer error, see e.g. Ebenbauer et al. [39]. Here, we focus the attention on one problem which appears due to the geometry of the state space, i.e. the necessary existence of several equilibria for every continuous vector field with an asymptotically stable equilibrium.

As already mentioned, the state space of the control system considered here is not $\mathbb{R}^n$, but a smooth manifold which is not diffeomorphic to $\mathbb{R}^n$. More precisely, the state space is $SE(n) = \mathbb{R}^n \times SO(n)$, where $SO(n) = \{\Theta \in \mathbb{R}^{n \times n} | \Theta^{-1} = \Theta^\mathsf{T}, \det \Theta = 1\}$ is the set of special orthogonal matrices. It is well known that continuous vector fields on $SO(n)$ for $n = 2k + 1$ with $k \in \mathbb{N}$ need to have multiple equilibria in the presence of an asymptotically stable equilibrium, see e.g. the textbooks Milnor [97], Guillemin and Pollack [51] or the article Koditschek [73]. In the case $n = 2k + 1$ with $k \in \mathbb{N}$ this can be inferred with the help of the Poincaré-Hopf Theorem. This theorem states that a smooth vector field on a compact, oriented manifold $\mathcal{M}$ with only finitely many equilibrium points has the property that the sum of the indices of the vector field equals the Euler characteristic of $\mathcal{M}$, see e.g. Guillemin and Pollack [51]. A compact, odd-dimensional, oriented manifold $\mathcal{M}$ (like $SO(2k + 1)$) has Euler characteristic zero [51, p.116]. An asymptotically stable equilibrium has a non-zero index, see Thews [144], hence a globally asymptotically stable equilibrium without the presence of additional equilibria is impossible.

Solving the output regulation problem includes a stabilization problem on that state space. In our case this means the stabilization of a point on $SE(n)$, and thus the stabilization of a point on $SO(n)$. Since we aim for smooth stabilization of an equilibrium, the considered closed loop systems are going to have multiple equilibria. As a consequence, any smooth stabilization approach cannot achieve global stability of a desired equilibrium. In such a situation, a desirable convergence behavior would be that the ω-limit set of every solution lies in one of finitely many isolated components of the equilibrium set, that there is only one asymptotically stable equilibrium and the other components of the equilibrium set are unstable. For the problems considered here, the ω-limit sets are not only equilibria, but isolated sets. The desirable convergence behavior in this situation is summarized in the following:

Definition 3.1. *Consider the differential equation $\dot{x} = f(x)$ with $x \in M \subset \mathbb{R}^n$ and a smooth $f : M \to \mathbb{R}^n$. Suppose there is a finite number of isolated sets $N_0, \ldots, N_l$ with $l \in \mathbb{N}$. Then we say that the solution of $\dot{x} = f(x)$ typically converges to N_0 if the following properties hold:*

a) *All solutions remain bounded and the ω-limit set of any solution is contained in exactly one set N_i for an $i \in \{0, \ldots, l\}$.*

b) *N_0 is asymptotically stable.*

c) *$N_1, \ldots, N_l$ are unstable.*

In the situation considered here, we show that the sets $N_0, \ldots, N_l$ are submanifolds of the considered state space and the dimension of these submanifolds is lower than the state space dimension. Suppose that the state x of $\dot{x} = f(x)$ is partitioned into $x = (x_1, x_2) \in M_1 \times M_2$ and $N_0 = \{x | x_1 = d\}$ with $d \in M_1$. Then we use the notation

$$\lim_{t \to \infty} x_1(t) \xrightarrow{\text{typ}} d \tag{3.4}$$

if the solution of $\dot{x} = f(x)$ typically converges to N_0. A stricter requirement would be *almost global stability of the desired invariant set*. Loosely speaking, this means that the desired invariant set is asymptotically stable and all initial conditions except a "thin" set lead to solutions which converge to the desired invariant set. Almost global approaches for state feedback designs for rigid body problems and problems on $SE(n)$ are well known, see e.g. Koditschek [73] or Bhat and Bernstein [15].

However, the multiple equilibria cause an additional difficulty for the observer-based approaches. This difficulty presents a challenge to the typical closed loop convergence in the sense of Definition 3.1 and also for the almost global convergence. To explain the key problem, we consider a system of the form

$$\begin{aligned} \dot{x} &= f(x) \\ \dot{y} &= g(x, y), \end{aligned} \tag{3.5}$$

where the state $x \in \mathcal{N}_1$, the state $y \in \mathcal{N}_2$ and $\mathcal{N}_1, \mathcal{N}_2$ are smooth manifolds. We assume that the vector fields given by f, g are smooth. Assume that all solutions $t \mapsto (x(t; x_0), y(t; (x_0, y_0))$ remain bounded for all initial conditions $x_0 \in \mathcal{N}_1$, $y_0 \in \mathcal{N}_2$ and that $x(t; x_0) \to 0$ for $t \to \infty$ for all $x_0 \in \mathcal{N}_1$. Furthermore assume that every solution of the differential equation

$$\dot{y} = g(0, y) \tag{3.6}$$

converges to a single equilibrium in the set of equilibria, but there exist multiple possibly isolated equilibria. In other words, the ω-limit set $\omega((x_0,y_0))$ of a point $(x_0,y_0) \in \mathscr{N}_1 \times \mathscr{N}_2$ with $x_0 = 0$ is given by a single point $(0,y) \in \mathscr{N}_1 \times \mathscr{N}_2$. If $x(t;x_0) \to 0$ for $t \to \infty$ in (3.5) for all $x_0 \in \mathscr{N}_1$, then (3.6) would be the limit equation of (3.5). The question then is, whether the ω-limit sets of a point $(x_0,y_0) \in \mathscr{N}_1 \times \mathscr{N}_2$ with $x_0 \neq 0$ for (3.5) is then also given by a single point $(0,y) \in \mathscr{N}_1 \times \mathscr{N}_2$. We show in the Examples 3.2 and 3.3 that this is not necessarily the case.

Example 3.2. *Consider the system*

$$\dot{x} = -x^3$$
$$\dot{y}_1 = -xy_2 \tag{3.7}$$
$$\dot{y}_2 = xy_1.$$

Any solution $t \mapsto \big(x(t;(x_0,y_{1,0},y_{2,0})),y_1(t;(x_0,y_{1,0},y_{2,0})),y_2(t;(x_0,y_{1,0},y_{2,0}))\big)$ *of the system stays bounded, since* $(x,y_1,y_2) \mapsto \frac{1}{2}x^2 + y_1^2 + y_2^2$ *is a proper Lyapunov function. Furthermore, for any solution we have* $x(t;(x_0,y_{1,0},y_{2,0})) \to 0$ *for* $t \to \infty$*, i.e. all solutions converge towards the* y_1-y_2-plane $\mathscr{N} = \{x \in \mathbb{R}^3 | x_1 = 0\}$ *which is an embedded submanifold of* $\mathscr{M} = \mathbb{R}^3$*. Every point* $\big(0,y_{1,0},y_{2,0}\big)^\mathsf{T} \in \mathbb{R}^3$ *is an equilibrium point of (3.7), therefore the* ω-limit set of $\big(0,y_{1,0},y_{2,0}\big)^\mathsf{T} \in \mathbb{R}^3$ *with is* $\big(0,y_{1,0},y_{2,0}\big)^\mathsf{T}$*.*

In the case of an initial condition $\big(x_0,y_{1,0},y_{2,0}\big)^\mathsf{T} \in \mathbb{R}^3$ *with* $x_0 \neq 0$*, it can be shown that its* ω-limit set is a circle lying in the y_1-y_2-plane. To see this, let $x_0 \neq 0$ be arbitrary but fixed. We integrate the first equation in (3.7), which is independent of the second and the third equation and hence of $y_{1,0},y_{2,0}$, and define a function $\phi : \mathbb{R}_{\geq 0} \to \mathbb{R}$ by $\phi(t) = x(t;x_0)$. Let $t \mapsto \big(z_1(t;0,(y_{1,0},y_{2,0})),z_2(t;0,(y_{1,0},y_{2,0}))\big)$ denote a solution to

$$\dot{z}_1 = -\phi(t)z_2$$
$$\dot{z}_2 = \phi(t)z_1 \tag{3.8}$$

with the initial condition $\big(y_{1,0},y_{2,0}\big) \in \mathbb{R}^2$ *and initial time* $t_0 = 0$*. Then the function defined by* $t \mapsto \big(\phi(t),z_1(t;0,(y_{1,0},y_{2,0})),z_2(t;0,(y_{1,0},y_{2,0}))\big)$ *is the solution to (3.7) with* $\big(x_0,y_{1,0},y_{2,0}\big)^\mathsf{T}$ *as initial condition. Since*

$$\frac{d}{dt}\left((z_1(t;0,(y_{1,0},y_{2,0})))^2 + (z_2(t;0,(y_{1,0},y_{2,0})))^2\right)$$
$$= 2z_1(t;0,(y_{1,0},y_{2,0}))2\dot{z}_1(t;0,(y_{1,0},y_{2,0})) + 2z_2(t;0,(y_{1,0},y_{2,0}))2\dot{z}_2(t;0,(y_{1,0},y_{2,0})) \tag{3.9}$$
$$= 0.$$

we see that the solutions fulfill $(z_1(t;0,(y_{1,0},y_{2,0})))^2 + (z_2(t;0,(y_{1,0},y_{2,0})))^2 = \big|(y_{1,0},y_{2,0})^\mathsf{T}\big|^2$ *for all* $t \geq 0$*, which means that the orbit of a solution of (3.7) with* $x_0 \neq 0$ *projected on the* y_1-y_2-plane is a circle.

From Example 3.2 one could get the impression that the reason for the problem is that every point in the y_1-y_2-plane is an equilibrium for (3.7). This is not the case as the following example shows.

Example 3.3 (Thieme [145, Example 1.5]). *Consider the system*

$$\dot{r} = r(1-r)$$
$$\dot{\phi} = \beta r|\sin(\phi)| + x \tag{3.10}$$
$$\dot{x} = -\gamma x,$$

where the states r, ϕ, x *fulfill* $r \in \mathbb{R}_{>0}$, $\phi \in [0, 2\pi)$ *and* $x \in \mathbb{R}$ *and the parameters* $\beta, \gamma \in \mathbb{R}$ *fulfill* $\beta > 0$ *and* $\gamma > 0$.

The equilibria $(\bar{r}, \bar{\phi}, \bar{x})$ *of the system are determined by setting the right hand side of the vector field given by (3.7) equal to zero. This directly yields* $\bar{r} = 1$ *and* $\bar{x} = 0$. *From* $\bar{\phi} \in [0, 2\pi)$ *and* $\sin(\bar{\phi}) = 0$ *we thus get either* $\bar{\phi} = 0$ *or* $\bar{\phi} = \pi$. *Therefore the equilibria of the system are* $(1, 0, 0)$ *and* $(1, \pi, 0)$.

The r-*equation is independent and* $\bar{r} = 1$ *is the only equilibrium of the* r-*equation. Utilizing the Lyapunov function* $r \mapsto (r - 1)^2$ *shows that all solutions with* $r_0 \in \mathbb{R}_{>0}$ *converge to* $\bar{r} = 1$. *If the initial condition* x_0 *is zero, then the solutions* $t \mapsto (r(t; r_0), \phi(t; (r_0, \phi_0, 0), x(t; 0))$ *converge either to* $(1, 0, 0)$ *or to* $(1, \pi, 0)$.

If the initial condition x_0 *for the* x-*equation is non-zero, then the solution component* $t \mapsto \phi(t; (r_0, \phi_0, x_0))$ *is unbounded, which means that* $t \mapsto \phi(t; (r_0, \phi_0, x_0))$ *takes every value in* $[0, 2\pi)$ *and thus the* ω-*limit set of such solutions is the whole circle with radius* 1. *For details see Thieme [145].*

From a control point of view the x-part in (3.5) would be the observer error which converges for all initial conditions to zero. The dynamics (3.6), i.e. the y-part in (3.6) for $x = 0$, would be the dynamics of the plant together with the state feedback. The problem in a control language would then be as explained in the following. Despite the boundedness of the closed loop and the convergence of the observer error for all initial conditions, we cannot in general guarantee the convergence to the desired equilibrium for the solutions in the closed loop consisting of the plant and a certainty equivalence implementation of the state feedback. The reason for the problem are the multiple isolated equilibria in the dynamics (3.6), i.e. in the dynamics of the plant together with the state feedback.

In this thesis, we present an approach where we can guarantee typical convergence to the desired equilibrium in the sense of Definition 3.1 for the certainty equivalence implementation of the feedback despite the presence of multiple equilibria. Our approach utilizes that, in many cases, asymptotic stability of the desired invariant set can be shown in the state feedback case with the help of an auxiliary function, i.e. a Lyapunov-like function. A question which is relevant for output regulation problems is whether the information provided by the auxiliary function can be utilized to circumvent the problems described above and establish the typical convergence in the closed loop for the output feedback case. In other words, given a state feedback which achieves typical convergence for the closed loop and it possible to show these properties with the help of an auxiliary function, is it possible to show that a certainty equivalence implementation of the feedback with a typically convergent observer is a suitable output feedback? In this chapter, we give an affirmative answer to this question for control problems on $SE(n)$ and rigid body control problems in the sense that we can guarantee the typical convergence in the closed loop.

The presented method is not restricted to the aforementioned system classes. If a suitable auxiliary function exists, then the proposed design can be adapted to other problems and is going to exhibit a similar separation property.

3.2   Global output regulation for control systems on $SE(n)$

In this section, we are going to present a global solution to a class of nonlinear output regulation problems for systems on the special Euclidean group $SE(n)$. The special Euclidean group $SE(n)$ describes translations and rotations of the n-dimensional Euclidean space and is given by the product of $\mathbb{R}^n$ and the manifold of special orthogonal matrices $SO(n) = \{\Theta \in \mathbb{R}^{n \times n} | \Theta^\top \Theta = I, \det(\Theta) = 1\}$. The contributions of this section are summarized in the following. We present a controller design method for the considered class of output regulation problems. The presented solution is global and results in a smooth control law. We show that our approach achieves typical convergence to the desired equilibrium in the sense of Definition 3.1 and global boundedness of the closed loop solutions. The proposed solution is a two-step design method comprised of a state feedback design and an observer design. Despite the presence of multiple isolated equilibria in the observer error dynamics and in the state feedback loop, we show that the proposed design method achieves typical convergence and global boundedness of closed loop solutions. Since the design of the observer and the state feedback are independent, we establish a novel nonlinear separation principle for the considered system class. The results of this section are going to be published in Schmidt et al. [123]. Preliminary results of this section have been published in Schmidt et al. [120].

The remainder of the Section is structured as explained in the following. We give a detailed problem statement in Section 3.2.1. In Section 3.2.2, we present our state feedback design for the considered problem class. We design the observer for the states that are not available from measurements in Section 3.2.3. In Section 3.2.4 we establish the stability of the closed loop which uses a certainty equivalence implementation of the state feedback. We present an example in Section 3.2.5. Finally, we give a summary in Section 3.2.6.

3.2.1 Problem statement

In the following we describe the problem setup for the output regulation problem on $SE(n)$, i.e. we introduce the control system and the exosystem. Thereby we denote the state of the exosystem by $\mathcal{M}_1$ and the state space of the control system by $\mathcal{M}_2$. We use the subscript "1" for the states of the exosystem and the subscript "2" for the states of the control system. This is consistent with the subscripts of the corresponding state spaces. The state space $\mathcal{M}$ of the closed loop system then consists of the state space of the exosystem $\mathcal{M}_1$, the state space of the control system $\mathcal{M}_2$ and the state space of the controller $\mathcal{M}_3$ which we introduce in Section 3.2.3 and Section 3.2.4 respectively.

The class of control systems considered in this work has the form

$$
\begin{aligned}
\dot{b}_2 &= A_2 b_2 + p(w, b_2, \Theta_2) + u \\
\dot{\Theta}_2 &= (P(w, b_2, \Theta_2) + U)\Theta_2 \\
y_2 &= (b_2, \Theta_2),
\end{aligned}
\tag{3.11}
$$

where $(b_2, \Theta_2) \in SE(n)$, $A_2 \in \mathbb{R}^{n \times n}$, $P : \mathbb{R}^m \times SE(n) \to \mathbb{R}^{n \times n}$, $P(w, b_2, \Theta_2) = -P^\top(w, b_2, \Theta_2)$, $p : \mathbb{R}^m \times SE(n) \to \mathbb{R}^n$, u and U denote the control inputs, $SE(n) = \mathbb{R}^n \times SO(n)$ and $SO(n) = \{\Theta \in \mathbb{R}^{n \times n} | \Theta^{-1} = \Theta^\top, \det(\Theta) = 1\}$. P and p are possible nonlinearities acting on the control system. We assume that p is bounded with respect to the second argument, i.e. for any w and Θ_2 there is a $\delta_1, \delta_2 > 0$ such that $\delta_1 \le p(w, b_2, \Theta_2) \le \delta_2$ for all b_2. We denote the state space of the control system (3.11) by $\mathcal{M}_2$, i.e.

$$
\mathcal{M}_2 = \mathbb{R}^n \times SO(n).
\tag{3.12}
$$

The type of exosystem we consider is motivated by tracking applications, i.e. we assume that a part of the exosystem is a system on $SE(n)$ similar to the control system (3.11). We assume that the exosystem is of the form

$$\dot{b}_1 = A_1 b_1 + v(y_1, w)$$
$$\dot{\Theta}_1 = Q(w)\Theta_1 \qquad (3.13)$$
$$y_1 = (C_1 b_1, \Theta_1),$$

where $w \in \mathbb{R}^m$, $(b_1, \Theta_1) \in SE(n)$, $C_1 \in \mathbb{R}^{q \times n}$, $Q : \mathbb{R}^m \to T_I SO(n)$ is linear, $v : \mathbb{R}^q \times SO(n) \times \mathbb{R}^m \to \mathbb{R}^n$ is smooth. w is assumed to be a solution of the differential equation

$$\dot{w} = Sw \qquad (3.14)$$

with $S \in \mathbb{R}^{m \times m}$ and $S = -S^\mathsf{T}$. We assume that A_1 is Hurwitz, i.e. the eigenvalues of A_1 have a negative real part. Since $S = -S^\mathsf{T}$, the solutions of (3.14) are bounded. Furthermore, since A_1 is Hurwitz and v is bounded, the b_1-component of the solutions of (3.13) remains bounded as well. We denote the state space of the exosystem (3.13) and (3.14) by $\mathcal{M}_1$, i.e.

$$\mathcal{M}_1 = \mathbb{R}^m \times \mathbb{R}^n \times SO(n). \qquad (3.15)$$

The goal is that (b_2, Θ_2) asymptotically converges to (b_1, Θ_1). As a consequence the regulated output is the error between the actual and the desired state. For the b-part the usual choice for the error function is $e = b_2 - b_1$, for the Θ-part one usual choice is $E = \Theta_2 \Theta_1^{-1}$, see e.g. Wen and Kreutz-Delgado [147], Lageman et al. [77]. If $E = I$, then we have $\Theta_2 = \Theta_1$, as desired in tracking applications. Overall, the regulated output is given by

$$(e, E) = (b_2 - b_1, \Theta_2 \Theta_1^{-1}) \in SE(n). \qquad (3.16)$$

The goal is to achieve

$$(e(t), E(t)) = (b_2(t) - b_1(t), \Theta_2(t)\Theta_1^{-1}(t)) \to (0, I) \qquad (3.17)$$

for $t \to \infty$. The controller is assumed to be of the form

$$\dot{x}_c = F_c(x_c, y_1, y_2, E)$$
$$u = h_c(x_c, y_1, y_2, E) \qquad (3.18)$$
$$U = H_c(x_c, y_1, y_2, E),$$

where x_c is the state of the controller and F_c, h_c, H_c are assumed to be smooth. Note that we formally list E in the arguments of the vector fields in (3.18), ignoring the redundancy given by Θ_1, Θ_2. The complete setup is illustrated in Figure 3.1.

In summary, the considered regulation problem can be formulated as follows:

Problem statement 3.4. *Find a controller of the form* (3.18) *such that the closed loop* (3.11), (3.13), (3.14) *and* (3.18) *has the following properties:*

a) *The closed loop achieves*

$$\lim_{t \to \infty} (e(t), E(t)) \xrightarrow{typ} (0, I) \qquad (3.19)$$

in the sense of Definition 3.1.

b) *The zero error manifold* Δ, *defined by*

$$\Delta = \{(b_1, \Theta_1, b_2, \Theta_2) | b_2 = b_1, \ \Theta_2 \Theta_1^{-1} = I\}, \qquad (3.20)$$

is positively invariant for the closed loop system.

c) *All solutions of the closed loop system remain bounded.*

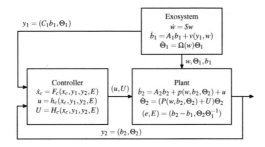

Figure 3.1: Illustration of the output regulation setup.

3.2.2 State feedback design

First, we derive a necessary condition for the state feedback that guarantees the invariance of the zero error manifold Δ given in (3.20).

Lemma 3.5. *Assume a controller of the form* (3.18) *solves the regulation problem as stated in Problem statement 3.4. If* $b_1 = b_2$, $\Theta_1 = \Theta_2$, *then the feedback* (U, u) *must have the form*

$$u = A_1 b_1 - A_2 b_2 - p(w, b_2, \Theta_2) + v(y_1, w)$$
$$U = -P(w, b_2, \Theta_2) + Q(w). \tag{3.21}$$

Proof. A controller which solves Problem statement 3.4 renders Δ defined in (3.20) invariant, i.e. for all $(b_1(0), \Theta_1(0), b_2(0), \Theta_2(0)) \in \Delta$ the condition $(e(t), E(t)) = (0, I)$ holds for all $t \geq 0$. This implies $(\dot{e}(t), \dot{E}(t)) = 0$ for all $t \geq 0$, i.e.

$$0 = \dot{e} = A_2 b_2 - A_1 b_1 + p(w, b_2, \Theta_2) - v(y_1, w) + u, \tag{3.22}$$
$$0 = \dot{E} = \dot{\Theta}_2 \Theta_1^{-1} - \Theta_2 \Theta_1^{-1} \dot{\Theta}_1 \Theta_1^{-1}$$
$$= (P(w, b_2, \Theta_2) + U) \Theta_2 \Theta_1^{-1} - \Theta_2 \Theta_1^{-1} Q(w) \tag{3.23}$$
$$= P(w, b_2, \Theta_2) + U - Q(w).$$

Hence (3.21) must hold. ∎

Equation (3.21) shows that the feedback needs w, b_1, Θ_1 in order to achieve invariance of the zero error manifold Δ. This can be achieved by including a system of the form (3.13) and (3.14) into the controller, whose states $\hat{w}, \hat{b}_1, \hat{\Theta}_1$ asymptotically reconstruct w, b_1, Θ_1.

Next we assume that we can measure the state of the exosystem. To present a solution to Problem statement 3.4, we construct a feedback (u, U) which depends on the states of the system (3.11) and the states of the exosystem (3.13), (3.14), and which almost globally stabilizes $(e, E) = (0, I)$. The control system (3.11), (3.13) and (3.14) written in error states (3.16) is given by

$$\dot{w} = Sw$$
$$\dot{b}_1 = A_1 b_1 + v(y_1, w)$$
$$\dot{\Theta}_1 = Q(w) \Theta_1 \tag{3.24}$$
$$\dot{e} = A_2 e + (A_2 - A_1) b_1 + p(w, b_1 + e, \Theta_2) - v(y_1, w) + u$$
$$\dot{E} = \left(P(w, b_2, \Theta_2) - E Q(w) E^{\mathsf{T}} \right) E + UE.$$

In order to stabilize $(e,E) = (0,I)$ we propose the feedback

$$u = -(I+A_2)e - (A_2 - A_1)b_1 - p(w, b_1 + e, \Theta_2) + v(y_1, w)$$
$$U = -P(w, b_2, \Theta_2) + EQ(w)E^{\mathsf{T}} + E^{\mathsf{T}} - E. \tag{3.25}$$

The resulting (state feedback) closed loop system is then given by the first three equations in (3.24) and

$$\dot{e} = -e$$
$$\dot{E} = (E^{\mathsf{T}} - E)E. \tag{3.26}$$

The following lemma summarizes the stability properties of (3.26):

Lemma 3.6. *Consider the system* (3.26).

a) *The ω-limit set of any solution is contained in the set of equilibria.*

b) *The equilibrium $(0,I)$ is locally exponentially stable and all other equilibria are unstable.*

c) *The solutions of* (3.26) *converge for almost all initial conditions towards $(0,I)$.*

Proof. **a)** The equilibria of (3.26) are given by $\{0\} \times \{E \in SO(n) | E = E^{\mathsf{T}}\}$. Consider the function $V_1 : \mathcal{M}_1 \to \mathbb{R}$ defined by

$$V_1(e,E) = \frac{1}{2}e^{\mathsf{T}}e + n - \operatorname{tr}(E). \tag{3.27}$$

If A is skew-symmetric, then $\operatorname{tr}(A^2) = \sum_{l=1}^{n}(A^2)_{ll} = -\sum_{l=1}^{n}\sum_{k=1}^{n}A_{lk}A_{kl} = -\sum_{l=1}^{n}\sum_{k=1}^{n}A_{lk}^2 \leq 0$. The derivative of V_1 along (3.26) fulfills

$$\dot{V}_1 = e^{\mathsf{T}}\dot{e} - \operatorname{tr}(\dot{E})$$
$$= -e^{\mathsf{T}}e - \operatorname{tr}\left((E^{\mathsf{T}} - E)E\right)$$
$$= -e^{\mathsf{T}}e - \frac{1}{2}\operatorname{tr}\left((E^{\mathsf{T}} - E)E + E^{\mathsf{T}}(E - E^{\mathsf{T}})\right) \tag{3.28}$$
$$= -e^{\mathsf{T}}e - \frac{1}{2}\operatorname{tr}\left(-EE + 2E^{\mathsf{T}}E - E^{\mathsf{T}}E^{\mathsf{T}}\right)$$
$$= -e^{\mathsf{T}}e + \frac{1}{2}\operatorname{tr}\left((E - E^{\mathsf{T}})^2\right) \leq 0.$$

As a consequence, the solutions of (3.26) converge to the set of equilibria of (3.26), i.e. the ω-limit set of the set of solutions is contained in the set of equilibria.

b) Since the dynamics of $\dot{e} = -e$ and $\dot{E} - (E^{\mathsf{T}} - E)E$ are not coupled, the stability properties of the equilibria follow from the stability properties of the equilibrium 0 for $\dot{e} = -e$ and from the stability properties of the equilibria $\{E \in SO(n) | E = E^{\mathsf{T}}\}$ for $\dot{E} = (E^{\mathsf{T}} - E)E$.

$\dot{e} = -e$ is a linear differential equation with the system matrix $-I$ with only one equilibrium at 0. Since the system matrix has n eigenvalues with negative real part, this equilibrium is globally asymptotically stable. Hence the ω-limit set of any solution of $\dot{e} = -e$ is the equilibrium 0. Since $\dot{e} = -e$ is linear and the real parts of the eigenvalues are negative, 0 is exponentially stable.

The stability of the equilibria $\{E \in SO(n) | E = E^{\mathsf{T}}\}$ for $\dot{E} = (E^{\mathsf{T}} - E)E$ are given in Lemma 3.30 in Section 3.4. Hence the identity matrix is the only exponentially stable equilibrium, the other equilibria are unstable. Hence b) follows.

c) Since the dynamics of $\dot{e} = -e$ and $\dot{E} = (E^{\mathsf{T}} - E)E$ are not coupled, the convergence properties of $\dot{e} = -e$ and of $\dot{E} = (E^{\mathsf{T}} - E)E$ are independent. The equilibrium 0 for $\dot{e} = -e$ is globally asymptotically stable. According to Corollary 3.31, the identity matrix I is almost globally asymptotically stable for $\dot{E} = (E^{\mathsf{T}} - E)E$. Hence c) follows. ∎

Lemma 3.7. *Assume that A_1 is Hurwitz. Then the control system* (3.24) *together with the state feedback* (3.25) *has the properties 3.4.a), 3.4.b) and 3.4.c).*

Proof. The state feedback closed loop dynamics consist of the first three equations in (3.24) and (3.26). The dynamics of the error (e, E) given by (3.26) are decoupled from the exosystem dynamics. Furthermore, the convergence properties of the error (e, E) are given in Lemma 3.6, i.e. for almost all initial conditions the error (e, E) converges asymptotically to $(0, I)$. Therefore property 3.4.a) holds. Moreover, if $(e, E) = (0, I)$, then (3.25) is

$$
\begin{aligned}
u &= -(I + A_2)e - (A_2 - A_1)b_1 - p(w, b_2, \Theta_2) + v(y_1, w) \\
&= (A_1 - A_2)b_1 - p(w, b_2, \Theta_2) + v(y_1, w). \\
U &= P(w, b_2, \Theta_2) + EQ(w)E^{\mathsf{T}} + E^{\mathsf{T}} - E \\
&= P(w, b_2, \Theta_2) + Q(w).
\end{aligned}
\tag{3.29}
$$

Hence, the feedback leaves Δ positively invariant and thus the property 3.4.b) holds. The remaining point is to show that the states of the closed loop remain bounded. The states of the closed loop consist of the exosystem components (w, b_1, Θ_1) and the control system components (b_2, Θ_2). Since the exosystem dynamics given by (3.13) and (3.14) are independent of the control system dynamics, the boundedness properties of the respective closed loop components remain unchanged. More precisely, the w and the b_1-component of the closed loop solutions remain bounded as explained below equation (3.13). The Θ_1- and the Θ_2-component of the solution are bounded since $SO(n)$ is compact. Since $e = b_2 - b_1$ we have $|b_2| \leq |b_1| + |e|$. Since the b_1-component of the solutions remains bounded and the e-component of the solution of the error dynamics remains bounded as well, the b_2-component of the solution of the closed loop remains bounded. Consequently all solutions of the closed loop remain bounded and therefore property 3.4.c) holds. ∎

3.2.3 Observer design

In this section we design an observer for the states of the exosystem. In Section 3.2.4, the observer will be used in combination with the feedback designed in Section 3.2.2. The observer reconstructs (b_1, w) from $(C_1 b_1, \Theta_1)$-measurements. We consider the systems (3.13) and (3.14) given by

$$
\begin{aligned}
\dot{w} &= Sw \\
\dot{b}_1 &= A_1 b_1 + v(y_1, w) \\
\dot{\Theta}_1 &= Q(w)\Theta_1 \\
y_1 &= (C_1 b_1, \Theta_1),
\end{aligned}
\tag{3.30}
$$

where $w \in \mathbb{R}^m$, $(b_1, \Theta_1) \in SE(n)$, $v : \mathbb{R}^q \times SO(n) \times \mathbb{R}^m \to \mathbb{R}^n$ is smooth and we assume that $Q : \mathbb{R}^m \to T_I(SO(n))$ is of the form

$$
Q(w) = \sum_{i=1}^{m} Q_i w_i
\tag{3.31}
$$

with linearly independent Q_i. System (3.30) corresponds to the exosystem. The goal is to design an observer that allows to recover w and b_1 from the output y_1. We define the errors as $e_w = \hat{w} - w$, $e_b = \hat{b}_1 - b_1$ and $E_\Theta = \hat{\Theta}_1 \Theta_1^{-1}$ and use a Luenberger-type observer structure, i.e.

$$\begin{aligned}
\dot{\hat{w}} &= S\hat{w} + L_w(y_1, \hat{b}_1, \hat{\Theta}_1) \\
\dot{\hat{b}}_1 &= A_1\hat{b}_1 + v(y_1, \hat{w}) + L_b(y_1, \hat{b}_1, \hat{\Theta}_1)C_1 e_b \\
\dot{\hat{\Theta}}_1 &= \left(Q(\hat{w}) + L_\Theta(y_1, \hat{b}_1, \hat{\Theta}_1) \right)\hat{\Theta}_1 \\
\hat{y}_1 &= (C_1\hat{b}_1, \hat{\Theta}_1),
\end{aligned} \tag{3.32}$$

where L_w, L_b, L_Θ are nonlinear observer gains. We denote the state space of (3.32) by

$$\mathscr{M}_3 = \mathbb{R}^m \times \mathbb{R}^n \times SO(n). \tag{3.33}$$

The error dynamics are given by

$$\begin{aligned}
\dot{e}_w &= S e_w + L_w(y_1, \hat{b}_1, \hat{\Theta}_1) \\
\dot{e}_b &= A_1 e_b + (v(y_1, \hat{w}) - v(y_1, w)) + L_b(y_1, \hat{b}_1, \hat{\Theta}_1)C_1 e_b \\
\dot{E}_\Theta &= \dot{\hat{\Theta}}_1 \Theta_1^{-1} - \hat{\Theta}_1 \Theta_1^{-1} \dot{\Theta}_1 \Theta_1^{-1} \\
&= \left(Q(\hat{w}) + L_\Theta(y_1, \hat{b}_1, \hat{\Theta}_1) \right) E_\Theta - E_\Theta Q(w) \\
&= \left(Q(e_w) + L_\Theta(y_1, \hat{b}_1, \hat{\Theta}_1) \right) E_\Theta + [Q(w), E_\Theta].
\end{aligned} \tag{3.34}$$

In addition to the observer structure, we need an observability assumption. More specifically, we have to make an assumption on the influence of e_w on the (e_b, E_Θ)-dynamics. Loosely speaking, we need to have the property that as long as the e_w dynamics do not converge to zero asymptotically, e_b does not converge to zero and E_Θ does not converge to the unit matrix. Since e_w can be considered as the input of the (e_b, E_Θ)-system, this assumption is a special case of *output-input stability* as presented in Liberzon et al. [84]. Another interpretation is that we assume an asymptotic *zero-input detectability property* as defined in Sepulchre et al. [128]. Formally, we define:

Definition 3.8. *Let $\dot{x} = f(x, w)$ be a system with output $y = h(x)$. We call the output y of such a control system asymptotically w-detectable if $y(t) \to 0$ for $t \to \infty$ implies $w(t) \to 0$ for $t \to \infty$.*

Lemma 3.9. *Consider the system given by (3.13) and (3.14) (exosystem) together with the observer error dynamics (3.34). Assume there are smooth observer gains L_w, L_b, L such that the solution components $t \mapsto e_w(t; (x_0, \hat{x}_0))$, $t \mapsto e_b(t; (x_0, \hat{x}_0))$ of the system (3.13), (3.14) with (3.34) remain bounded for all initial conditions $x_0 = (w_0, b_{1,0}, \Theta_{1,0}) \in \mathscr{M}_1$ and $\hat{x}_0 = (e_{w,0}, e_{b,0}, E_{\Theta,0}) \in \mathscr{M}_3$. Furthermore assume that $L_\Theta(y_1, \hat{b}_1, \hat{\Theta}_1) = 0$ if $e_b = 0$ and $E_\Theta = I$ and that L_Θ is smooth. Then, the output $y = (e_b, E_\Theta - I)$ is asymptotically e_w-detectable.*

Proof. To avoid a too cluttered notation, we drop for this proof the explicit dependence of the solution components on the initial conditions. More precisely, we write $t \mapsto e_b(t)$ for the e_b-component of the solution of (3.34), and likewise for the e_w and the E_Θ-component. Assume that $(e_b(t), E_\Theta(t)) \to (0, I)$ for $t \to \infty$. Since $E_\Theta(t) = E_\Theta(0) + \int_0^t \dot{E}_\Theta(\tau)d\tau$, we know that

$$\lim_{t \to \infty} \int_0^t \dot{E}_\Theta(\tau)d\tau = I - E_\Theta(0). \tag{3.35}$$

By assumption, the e_w and the e_b-component of the solution of the error dynamics remain bounded. The compactness of $SO(n)$ implies that the E_Θ-component is bounded. Differentiation of $t \mapsto \dot{E}_\Theta(t)$ reveals that $t \mapsto \ddot{E}_\Theta(t)$ remains bounded. Since $t \mapsto \ddot{E}_\Theta(t)$ remains bounded and is continuous, $t \mapsto \dot{E}_\Theta(t)$ is Lipschitz, which implies that $t \mapsto \dot{E}_\Theta(t)$ is uniformly continuous, see e.g. Amann and Escher [4, Chapter VI.2]. Since $t \mapsto \dot{E}_\Theta(t)$ is uniformly continuous and $\lim_{t\to\infty} \int_0^t \dot{E}_\Theta(\tau)d\tau$ exists and is finite, Barbalat's Lemma implies $\dot{E}_\Theta(t) \to 0$ for $t \to \infty$, see e.g. Krstić et al. [75, Lemma A.6]. Consequently, the right hand side of the third equation in the error dynamics (3.34) must fulfill

$$\lim_{t\to\infty} [E_\Theta(t), Q(w(t))] - \big(Q(e_w(t)) + L_\Theta(y_1, \hat{b}_1, \hat{\Theta}_1)\big) E_\Theta(t) = 0. \tag{3.36}$$

Since $e_b(t) = 0$ and $E_\Theta(t) = I$ for $t \to \infty$ and since L_Θ is smooth, we have $Q(e_w(t)) = 0$ for $t \to \infty$. Because of (3.31), $Q(e_w)$ is linear in e_w and the Q_i are linearly independent, which implies $e_w(t) = 0$ for $t \to \infty$. ∎

Lemma 3.10. *Consider the system* (3.30) *together with the observer* (3.32) *and assume* (A_1, C_1) *is observable. Choose any constant matrix* L_b *such that* $A_1 + L_b C_1$ *is Hurwitz and let*

$$L_\Theta(y_1, \hat{b}_1, \hat{\Theta}_1) = E_\Theta^{\mathsf{T}} - E_\Theta \tag{3.37}$$

$$\big(L_w(y_1, \hat{b}_1, \hat{\Theta}_1)\big)_s = \sum_{l=1}^n \sum_{k=1}^n (Q_s)_{lk} (E_\Theta)_{kl}. \tag{3.38}$$

Then $(e_w(t), e_b(t), E_\Theta(t)) \overset{typ}{\longrightarrow} (0, 0, I)$ *for* $t \to \infty$ *in the sense of Definition 3.1 .*

Proof. We investigate the error dynamics (3.34) with the given observer gains in two steps. For the first step, note that the gains (3.37) and (3.38) do not depend on $\hat{b}_1$. This implies that the dynamics of e_w and E_Θ do not depend on e_b. Hence we consider the convergence properties of e_w and E_Θ independently from e_b together with the first equation from (3.30), i.e. $\dot{w} = Sw$. Consider $V_2 : \mathbb{R}^m \times SO(n) \to \mathbb{R}$ defined by

$$V_2(e_w, E_\Theta) = \frac{1}{2} e_w^{\mathsf{T}} e_w + n - \text{tr}(E_\Theta). \tag{3.39}$$

Since $V_2(e_w, E_\Theta) > 0$ for all $E_\Theta \in SO(n) \setminus \{I\}$, $e_w \in \mathbb{R}^m \setminus \{0\}$ the function V_2 is a Lyapunov function candidate. The derivative $\dot{V}_2$ is given by

$$\begin{aligned} \dot{V}_2 = &- \text{tr}\big(L_\Theta(y_1, \hat{b}_1, \hat{\Theta}_1) E_\Theta + Q(e_w) E_\Theta\big) \\ &+ e_w^{\mathsf{T}} L_w(y_1, \hat{b}_1, \hat{\Theta}_1), \end{aligned} \tag{3.40}$$

where we utilized that $e_w^{\mathsf{T}} S e_w = 0$ since S is skew-symmetric. Furthermore, since

$$\begin{aligned} \text{tr}(Q(e_w) E_\Theta) &= \sum_{l=1}^n (Q(e_w) E_\Theta)_{ll} = \sum_{l=1}^n \sum_{k=1}^n (Q(e_w))_{lk} (E_\Theta)_{kl} \\ &= \sum_{s=1}^m (e_w)_s \sum_{l=1}^n \sum_{k=1}^n (Q_s)_{lk} (E_\Theta)_{kl}, \end{aligned} \tag{3.41}$$

we obtain

$$\dot{V}_2 = \text{tr}\big((E_\Theta - E_\Theta^{\mathsf{T}}) E_\Theta\big) = \frac{1}{2} \text{tr}\big((E_\Theta - E_\Theta^{\mathsf{T}})^2\big) \leq 0. \tag{3.42}$$

Hence we can guarantee the convergence of E_Θ towards the set of equilibria given by $\{E_\Theta \in SO(n)|E_\Theta = E_\Theta^\top\}$. Furthermore, (3.42) implies that the e_w-component of the solution remains bounded. Therefore the sets containing the ω-limit sets for the (e_w, E_Θ)-dynamics and $\dot{w} = Sw$ are given by

$$N_0 = \{(w, e_w, E_\Theta) \in \mathbb{R}^m \times \mathbb{R}^m \times SO(n)|E_\Theta = I\} \text{ and}$$
$$N_k = \{(w, e_w, E_\Theta) \in \mathbb{R}^m \times \mathbb{R}^m \times SO(n)|E_\Theta = E_\Theta^\top, E_\Theta \neq I, \text{tr}(E_\Theta) = n - 4k\} \tag{3.43}$$

for $k \in \{1, \ldots, \lfloor \frac{n}{2} \rfloor\}$. For the instability of the N_k, fix one N_k and consider an initial condition $(w_0, e_{w,0}, E_{\Theta,0})$ where $e_{w,0} = 0$, $E_{\Theta,0}$ is arbitrarily close to an $E_\Theta = E_\Theta^\top$ with $(w_0, e_{w,0}, E_\Theta) \in N_k$ and $\text{tr}(E_{\Theta,0}) = c > n - 4k$. Since $V_2(e_{w,0}, E_{\Theta,0}) = n - \text{tr}(E_{\Theta,0}) = n - c < 4k$, $\dot{V}_2 \leq 0$ and the solutions are forward complete, we know $V_2(e_w(t), E_\Theta(t)) < 4k$ for all $t \geq 0$, where $e_w(t) = e_w(t; (e_{w,0}, E_{\Theta,0}))$ and $E_\Theta = E_\Theta(t; (w_0, e_{w,0}, E_{\Theta,0}))$. Next we show that this solution converges to one N_l with $l \neq k$. Suppose this is not the case, then $n - \text{tr}(E_\Theta(t)) = 4k$ for $t \to \infty$. On the other hand $V_2(e_w(t), E_\Theta(t)) = \frac{1}{2}e_w^\top(t)e_w(t) + n - \text{tr}(E_\Theta(t)) < 4k$, hence there would be a $t^* > 0$ such that $e_w^\top(t^*)e_w(t^*) < 0$, which is a contradiction. Finally, the components N_k are isolated, which implies the instability of N_k.

In case the E_Θ-component converges to the identity matrix, Lemma 3.9 implies that the e_w-component converges to zero. Thus N_0 is actually of the form $N_0 = \{(w, e_w, E_\Theta) \in \mathbb{R}^m \times \mathbb{R}^m \times SO(n)|e_w = 0, E_\Theta = I\}$. The asymptotic stability of the N_0 component follows from the properties that V_2 has a strict minimum in $(0, I)$ (see also Lemma 3.24 for a proof that $E_\Theta \mapsto n - \text{tr}(E_\Theta)$ has a strict minimum in I), that V_2 is positive definite with respect to the distance to N_0, that V_2 is bounded from above by a continuous and strictly increasing function of the distance to N_0, that $\dot{V}_2 < 0$ on a neighborhood of N_0 and the set stability theorem from Bhatia and Szegö [16, Chapter V, 4.18 Theorem].

In the second step we establish the convergence properties of the e_b-component of the solution of the error system (3.34). The observer error dynamics of e_b are given by

$$\dot{e}_b = (A_1 + L_b C_1)e_b + (v(y_1, \hat{w}) - v(y_1, w)), \tag{3.44}$$

i.e. an asymptotically stable linear system with bounded "input" $(v(y_1, \hat{w}) - v(y_1, w))$. Note that the w- and the b_1- component of the solutions of (3.14) remain bounded, as explained below equation (3.14), and that the $\hat{w}$-component of the observer remains bounded because of (3.42). As a consequence, the output $y_1 = (C_1 b_1, \Theta_1)$ is a function that remains bounded. Furthermore, the term $v(y_1, \hat{w}) - v(y_1, w)$ remains bounded. Therefore the e_b-component of the solution of the error system (3.34) remains bounded for all initial conditions. In the case that the solutions converges to N_0, i.e. $(e_w(t), E_\Theta(t)) \to (0, I)$ for $t \to \infty$, we have $v(y_1(t), \hat{w}(t)) - v(y_1(t), w(t)) \to 0$ for $t \to \infty$ and since $A_1 + L_b C_1$ is Hurwitz, the e_b-component of the solution of the error system (3.34) converges to zero. It remains to investigate the stability properties of the sets containing the ω-limit sets for the $(e_b, b_1, \Theta_1, w, e_w, E_\Theta)$-dynamics. These are either the set $\{0\} \times \mathbb{R}^n \times SO(n) \times N_0$ or subsets of $\mathbb{R}^n \times \mathbb{R}^n \times SO(n) \times N_k$. The instability of the subsets of $\mathbb{R}^n \times \mathbb{R}^n \times SO(n) \times N_k$ follows directly from the instability of the sets N_k for the (w, e_w, E_Θ)-dynamics, since the dynamics of e_w and E_Θ are independent of b_1, Θ_1 and e_b. The stability of the set $\{0\} \times \mathbb{R}^n \times SO(n) \times N_0$ follows from the fact that all solutions remain bounded, that the $(b_1, \Theta_1, w, e_w, E_\Theta)$-dynamics and the e_b-dynamics have a cascaded structure, i.e. the $(b_1, \Theta_1, w, e_w, E_\Theta)$-dynamics influence the e_b-dynamics, that $v(y_1, \hat{w}) - v(y_1, w)$ is smooth and zero on N_0, that $A_1 + L_b C_1$ is Hurwitz and that the solutions of the $(b_1, \Theta_1, w, e_w, E_\Theta)$-dynamics typically converge to $\mathbb{R}^n \times SO(n) \times N_0$. Using the property that N_0 is typically stable for the (w, e_w, E_Θ)-dynamics then implies the typical stability of $\{0\} \times \mathbb{R}^n \times SO(n) \times N_0$ for the $(e_b, b_1, \Theta_1, w, e_w, E_\Theta)$-dynamics using ISS cascade arguments. ∎

3.2.4 Output feedback design

In this section we use the feedback from Section 3.2.2 and the observer from Section 3.2.3 to solve the Problem 3.4 under the assumption that w is not available from measurements. We utilize a certainty equivalence controller. That means the closed loop consists of the system (3.11), the exosystem (3.13) and (3.14), the observer (3.32), and a certainty equivalence state feedback of the form

$$\begin{aligned}
u &= -(I + A_2)(b_2 - \hat{b}_1) - (A_2 - A_1)\hat{b}_1 \\
&\quad - p(\hat{w}, b_2, \Theta_2) + v(y_1, \hat{w}) \\
U &= -P(\hat{w}, b_2, \Theta_2) + EQ(\hat{w})E^\mathsf{T} + E^\mathsf{T} - E,
\end{aligned} \tag{3.45}$$

where $\hat{w}$ is the estimate of w, $\hat{b}_1$ is the estimate of b_1 and the observer gains L_b, L_w, L_Θ are chosen as in Lemma 3.10. The closed loop vector field in error states consists of the exosystem (3.13), (3.14), the observer error dynamics with error states e_w, e_b, E_Θ and the dynamics of e, E, i.e.

$$\begin{aligned}
\dot{e}_w &= Se_w + L_w(y_1, \hat{b}_1, \hat{\Theta}_1) \\
\dot{e}_b &= A_1 e_b + (v(\hat{y}_1, \hat{w}) - v(y_1, w)) + L_b(y_1, \hat{b}_1, \hat{\Theta}_1)C_1 e_b \\
\dot{E}_\Theta &= [E_\Theta, Q(w)] - (Q(e_w) + L_\Theta(y_1, \hat{b}_1, \hat{\Theta}_1))E_\Theta \\
\dot{e} &= A_2 e + (A_2 - A_1)b_1 - v(y_1, w) + p(w, b_1 + e, \Theta_2) + u \\
\dot{E} &= \left(P(w, b_2, \Theta_2) - EQ(w)E^\mathsf{T}\right)E + UE,
\end{aligned} \tag{3.46}$$

where u and U are given as in (3.45).

Before stating the result, we discuss the key difficulty to establish typical convergence for solutions of the closed loop in the sense of Definition 3.1. To solve Problem 3.4, we require that the invariant set where $(e_w, e_b, E_\Theta) = (0, 0, I)$ and $(e, E) = (0, I)$ is the only asymptotically stable set in the closed loop. Notice that $V_1 + V_2$ with V_1 from the proof of Lemma 3.6 and V_2 from the proof of Lemma 3.10 is not a Lyapunov function for the closed loop system, since the derivative $\dot{V}_1 + \dot{V}_2$ along solutions of the closed loop (3.13), (3.14), (3.45) and (3.46) contains indefinite expressions. Hence we cannot show convergence utilizing $V_1 + V_2$. Moreover, cascade arguments seem not to be applicable due to the presence of multiple disconnected sets of equilibria. Therefore, we choose another approach to show that the proposed certainty equivalence controller solves 3.4.

We utilize the general result on the asymptotic behavior of solutions of ordinary differential equations which approach an invariant manifold $\mathcal{N}$ from Arsie and Ebenbauer [9], which was repeated in Section 1.2.2. To apply this results here remember that the closed loop is given by (3.13), (3.14), (3.32), (3.45) and (3.11). In error states, the closed loop is given by (3.13), (3.14), (3.45) and (3.46). In our case, the closed loop in error states plays the role of the vector field f in the system $\dot{x} = f(x)$ for the height function. I.e., the state space $\mathcal{M}$ of the closed loop is given by

$$\mathcal{M} = \mathcal{M}_1 \times \mathcal{M}_2 \times \mathcal{M}_3, \tag{3.47}$$

where $\mathcal{M}_1 = \mathbb{R}^m \times SE(n)$ as defined in (3.15), $\mathcal{M}_2 = SE(n)$ as defined in (3.12) and $\mathcal{M}_3 = \mathbb{R}^m \times SE(n)$ as defined in (3.33). The state of the closed loop $x = (x_1, x_2, x_3)$ consists of the state of the exosystem $x_1 = (w, b_1, \Theta_1)$, the regulated output $x_2 = (e, E)$ and the error states of the observer $x_3 = (e_w, e_b, E_\Theta)$. The observer error dynamics are determined by the first three equations in (3.46) and the convergence of the observer errors $x_3 = (e_w, e_b, E_\Theta)$ is independent of the convergence of the error $x_2 = (e, E)$ since the right hand sides of the differential equations of $x_3 = (e_w, e_b, E_\Theta)$ do not depend on $x_2 = (e, E)$. Therefore Lemma 3.10 implies that for the observer error component $t \mapsto x_3(t; (x_1(0), x_2(0), x_3(0))$ of the solution of the closed loop typically converges towards $(0, 0, I)$, i.e. $x_3(t; (x_1(0), x_2(0), x_3(0))) \xrightarrow{\text{typ}} (0, 0, I)$ for $t \to \infty$ in the sense of Definition 3.1. Hence, the component

$$\mathcal{N} = \mathcal{M}_1 \times \mathcal{M}_2 \times (0, 0, I) \tag{3.48}$$

is the only asymptotically stable invariant set for the observer error. $\mathcal{N}$ is a embedded submanifold of $\mathcal{M}$. The dynamics on $\mathcal{N}$ are given by (3.24) and (3.25) and the asymptotic behavior of the dynamics on $\mathcal{N}$ is characterized in Lemma 3.6. More specifically, for initial conditions in $\mathcal{N}$, the solutions of the closed loop typically converge to $\mathcal{M}_1 \times (0, I) \times (0, 0, I)$. However, knowing the solutions typically converge to $\mathcal{N}$ asymptotically, does not imply in the general situation that the solutions typically converge to an equilibrium on $\mathcal{N}$, i.e. towards $\mathcal{E}$ given by

$$\mathcal{E} = \mathcal{M}_1 \times \mathcal{F} \times (0, 0, I)$$
$$\mathcal{F} = \{(e, E) \in SE(n) | e = 0, \ E^\mathsf{T} = E\}. \tag{3.49}$$

In other words, the asymptotic behavior of the dynamics on $\mathcal{N}$ does not necessarily describe the asymptotic behavior of solutions which are not initialized on $\mathcal{N}$. The reason for this are the multiple isolated equilibria of the dynamics on $\mathcal{N}$, for details see e.g. Arsie and Ebenbauer [9]. The situation is illustrated in Figure 3.2. In the following we show that the existence of

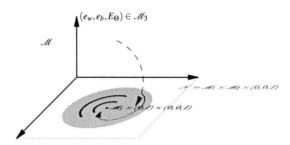

Figure 3.2: An illustration of the dynamics of the error system. The gray set illustrates $\mathcal{N}$. The black curves and the dot on $\mathcal{N}$ illustrate the connected components of equilibria. A solution which starts in $\mathcal{N}$ typically converges for towards the equilibrium $\mathcal{M}_1 \times (0, I) \times (0, 0, I)$ and typically the solutions converge towards $\mathcal{N}$. Since the restriction of the vector field to $\mathcal{N}$ has multiple equilibria, this does not imply typical convergence to the equilibrium in $\mathcal{N}$ or to any other equilibrium component in $\mathcal{N}$.

a suitable height function remedies the problem described above. The existence of a height function is established in the following:

Lemma 3.11. $V_3 : \mathscr{M} \to \mathbb{R}$ *defined by*

$$V_3(w, b_1, \Theta_1, e, E, e_w, e_b, E_\Theta) = \frac{1}{2} e^\mathsf{T} e + n - \operatorname{tr}(E) \tag{3.50}$$

is a height function for the pair $(\mathscr{N}, f)$ *where* $\mathscr{N}$ *is given by* (3.48) *and* f *is the vector field of the closed loop in error states given by* (3.13), (3.14), (3.45) *and* (3.46).

Proof. The domain of V_3 is $\mathscr{M}$, which is given by (3.47), but depends only on e and E. By Lemma 3.6, we have $\dot{V}_3|_{\mathscr{N}} \leq 0$ and $\dot{V}_3|_{\mathscr{N} \setminus \mathscr{E}} < 0$ where $\mathscr{E}$ is the set of $x \in \mathscr{N}$ where $\dot{V}_3$ vanishes. $\mathscr{E}$ is given by (3.49). ∎

Taking the previous results into account, we state now the main result of the section.

Theorem 3.12. *Assume that A_1 is Hurwitz. Then the controller consisting of the observer* (3.32) *where the observer gains* L_b, L_w, L_Θ *are chosen as in Lemma 3.10 together with the certainty equivalence feedback* (3.45) *solves Problem 3.4.*

Proof. The proof consists of three parts. In the first part we show that the conditions for the application of Theorem 1.17 are fulfilled. I.e. we show that solutions of the closed loop (3.13), (3.14), (3.32), (3.45) and (3.11) remain bounded, thereby proving that property 3.4.c) holds. In the second part we apply Theorem 1.17 to determine which sets can possibly contain the ω-limit set of the solutions which converge to $\mathscr{N}$. In the third part, we show that only one of these sets is stable while the others are unstable, thereby proving property 3.4.a). The property 3.4.b) holds because of Lemma 3.6.

First, we show that solutions of the closed loop remain bounded. The states of the closed loop are $x_1 = (w, b_1, \Theta_1)$, $x_2 = (b_2, \Theta_2)$ and $x_3 = (\hat{w}, \hat{b}_1, \hat{\Theta}_1)$. The w-, the b_1- and the Θ_1-component of the solution remain bounded since the dynamics of these components is given by the exosystem dynamics (3.13) and (3.14) and we assumed that the solutions of the exosystem remain bounded. More precisely, since $S = -S^\mathsf{T}$, the solutions of (3.14) remain bounded which implies that the w-component is bounded. Since A_1 is Hurwitz and v is bounded, the b_1-component remains bounded. The Θ_1-component is bounded since $\Theta_1 \in SO(n)$. For details, see also the explanations after (3.14). The observer error dynamics are determined by the first three equations in (3.46) and the convergence of the observer errors (e_w, e_b, E_Θ) is independent of the convergence of the error (e, E) since the right hand sides of the differential equations of (e_w, e_b, E_Θ) do not depend on (e, E). Therefore Lemma 3.10 guarantees that the observer error remains bounded. Because $\hat{w} = e_w + w$, $\hat{b}_1 = e_b + b_1$ and since the e_w-, the e_b-, the w- and the b_1-component of the solution remain bounded, the $\hat{w}$- and the $\hat{b}_1$-component remain bounded. All solution components which are element of $SO(n)$ are bounded, since $SO(n)$ is compact. Therefore the $\hat{\Theta}_1$- and the Θ_2-component are bounded. Since $e = b_2 - b_1$ and thus $|b_2| \leq |e| + |b_1|$, it is sufficient to show that the e-component of the closed loop error dynamics remains bounded to infer that the b_2-component of the closed loop dynamics remains bounded. To show that the e-component of the error dynamics remains bounded, we substitute the certainty equivalence feedback (3.45) into the $\dot{e}$-equation from (3.46). Rearranging the terms results in

$$\begin{aligned}
\dot{e} = -e - (I + A_1) e_b + v(y_1, \hat{w}) - v(y_1, w) \\
+ p(w, b_1 + e, \Theta_2) - p(\hat{w}, b_1 + e, \Theta_2),
\end{aligned} \tag{3.51}$$

i.e. the dynamics of e are $\dot{e} = -e + d$ with an exponentially stable part $\dot{e} = -e$ and an input $d = -(I + A_1) e_b + v(y_1, \hat{w}) - v(y_1, w) + p(w, b_2, \Theta_2) - p(\hat{w}, b_2, \Theta_2)$ given by the rest of the right hand side in (3.51). Since w, $\hat{w}$, $y_1 = (C_1 b_1, \Theta_1)$ and Θ_2 remain bounded and since p is bounded with respect to the second argument as explained below equation (3.11), $p(w, b_2, \Theta_2)$, $p(\hat{w}, b_2, \Theta_2)$ remain bounded. Hence, the input d of the system $\dot{e} = -e + d$ remains bounded and since $\dot{e} = -e$ is linear and has a unique asymptotically stable equilibrium at zero, we know that e remains bounded. As a consequence, property 3.4.c) is guaranteed.

In the second step we apply Theorem 1.17. Lemma 3.11 shows that V_3 defined by (3.50) is a height function for $(\mathcal{N}, f)$. The set $\mathscr{E}$ where $\dot{V}_3 = 0$ is given by (3.49), i.e. $\mathscr{E} = \mathscr{M}_1 \times \mathscr{F} \times (0, 0, I)$. Because of Lemma 3.24, we see that $\mathscr{F}$ are the critical points of the function $\Theta \mapsto n - \mathrm{tr}(\Theta)$ in Lemma 3.24. Consequently, the connected components $\mathscr{F}_k$ of $\mathscr{F}$ are given by

$$\mathscr{F}_k = \{(e, E) \in \mathbb{R}^n \times SO(n) | e = 0, \ E = E^\mathsf{T}, \mathrm{tr}(E) = n - 4k\} \tag{3.52}$$

for $k \in \{0, \ldots, \lfloor \frac{n}{2} \rfloor\}$. Since Lemma 3.24 shows that the trace is constant on the connected components of $\mathscr{F}$, we know that the connected components $\mathscr{E}_k = \mathscr{M}_1 \times \mathscr{F}_k \times (0, 0, I)$ of $\mathscr{E}$ are contained in level sets of $V_3|_{\mathcal{N}}$. The number of connected components of $\mathscr{E}$ is finite, hence the set $\{V_3(\mathscr{E}_k)\}_k$ is finite. Thus, according to Theorem 1.17, we know that the ω-limit set of a solution of (3.46) which converges to $\mathcal{N}$ lies in an unique $\mathscr{E}_k$.

In a final step we investigate the stability of the sets which possibly contain the ω-limit sets of the closed loop. First note that the stability properties of the invariant sets of the observer remain unchanged since the observer error dynamics do not depend on (e, E). Then Lemma 3.10 implies that the ω-limit sets are either subsets of $\mathscr{M}_1 \times \mathscr{M}_2 \times \mathscr{M}_3$ of the form

$$\{(w, b_1, \Theta_1, e, E, e_w, e_b, E_\Theta) \in \mathscr{M}_1 \times \mathscr{M}_2 \times \mathscr{M}_3 | E_\Theta = E_\Theta^\mathsf{T}, \ E_\Theta \neq I, \ \mathrm{tr}(E_\Theta) = n - 4p\} \tag{3.53}$$

for $p \in \{1, \ldots, \lfloor \frac{n}{2} \rfloor\}$ or of the form

$$\{(w, b_1, \Theta_1, e, E, e_w, e_b, E_\Theta) \in \mathscr{M}_1 \times \mathscr{M}_2 \times \mathscr{M}_3 | e_w = 0, \ e_b = 0, \ E_\Theta = I, \} . \tag{3.54}$$

In the case (3.53), the corresponding subsets of $\mathscr{M}_1 \times \mathscr{M}_2 \times \mathscr{M}_3$ are unstable because Lemma 3.10 implies that they are unstable in the observer case and the observer dynamics are independent of the (e, E)-dynamics. The only asymptotically stable subset for the observer dynamics is the set where $E_\Theta = I$, $e_w = 0$ and $e_b = 0$, i.e. (3.54). In the previous step of the proof we saw that in that case the ω-limit sets of the closed loop solutions are in a unique $\mathscr{E}_k = \mathscr{M}_1 \times \mathscr{F}_k \times (0, 0, I)$. Therefore, we investigate the stability of the connected components $\mathscr{E}_k$ of $\mathscr{E}$ which contain the ω-limit set $\omega(x_0)$ of a solution which converges to $\mathcal{N}$. The connected components $\mathscr{E}_k$ are characterized by Lemma 3.24, i.e. $\mathscr{E}_k = \mathscr{M}_1 \times \mathscr{F}_k \times (0, 0, I)$ where $\mathscr{F}_k$ is given in (3.52) and in Lemma 3.24c). Recall that the $\mathscr{F}_k$ correspond to the critical points of $(e, E) \mapsto \frac{1}{2} e^\mathsf{T} e + n - \mathrm{tr}(E)$ and to the equilibria of (3.26). It remains to show that the desired component $\mathscr{M}_1 \times (0, I) \times (0, 0, I)$ is the only asymptotically stable component and the other components are unstable.

To investigate the stability properties of the connected components, we consider the closed loop in error states (3.13), (3.14), (3.45) and (3.46) as a time-varying system consisting of (3.45) and (3.46) with w, b_1, Θ_1 as bounded time-varying functions.

To show the asymptotic stability of $(0, I) \times (0, 0, I)$, we utilize a Theorem by Iggidr and Sallet [62, Theorem 5] (see also Theorem 1.14), which allows the investigation of the stability of equilibria with semidefinite Lyapunov functions. Consider the positive semidefinite function $V_4 : \mathscr{M}_2 \times \mathscr{M}_3 \to \mathbb{R}$ defined by $V_4(e, E, e_w, e_b, E_\Theta) = \frac{1}{2} e_w^\mathsf{T} e_w + n - \mathrm{tr}(E_\Theta)$. A calculation utilizing the observer gains from Lemma 3.10 along the lines of the calculation in the proof of Schmidt et al. [120, Lemma 4] shows $\dot{V}_4 \leq 0$. As consequence, V_4 is positive semidefinite, $\dot{V}_4 \leq 0$ in a neighborhood of $(0, I) \times (0, 0, I)$ and $V_4 = 0$ if and only if $(e_w, E_\Theta) = (0, I)$. Now we want to consider the dynamics on the set where V_4 is zero, i.e. where the observer error $e_w = 0$ and $E_\Theta = I$. Since the observer gains L_b, L_w, L_Θ from Lemma 3.10 do not depend on $C_1 b_1$ and $\hat{b}_1$ and because of Lemma 3.10, this set is an invariant set. In other words, let $x_0 = (x_{0,1}, x_{0,2}, x_{0,3})$ denote an initial condition in the set where V_4 is zero, i.e. where $x_{0,1} = (w_0, b_{1,0}, \Theta_{1,0})$, $x_{0,2} = (b_{2,0}, \Theta_{2,0})$, $x_{0,3} = (\hat{w}_0, \hat{b}_{1,0}, \hat{\Theta}_{1,0})$, $w_0 = \hat{w}_0$ and $\Theta_{1,0} = \hat{\Theta}_{1,0}$. Then we have $e_w(t; x_0) = 0$ and $E_\Theta(t; x_0) = I$ for all $t \geq 0$, i.e. $w(t; x_0) = \hat{w}(t; x_0)$ and $\hat{\Theta}_1(t; x_0) = \Theta_1(t; x_0)$ for all $t \geq 0$. Hence the dynamics of (3.46) on the set where V_4 is zero are given by (3.46) with $\dot{e}_w(t; x_0) = 0$ and $\dot{E}_\Theta(t; x_0) = 0$ for all $t \geq 0$, i.e.

$$
\begin{aligned}
\dot{e}_b &= (A_1 + L_b C) e_b \\
\dot{e} &= A_2 e + (A_2 - A_1) b_1 + v(y_1, w) + p(w, b_1 + e, \Theta_2) + u \\
\dot{E} &= \left(P(w, b_2, \Theta_2) - E Q(w) E^{\mathsf{T}} \right) E + U E
\end{aligned}
\tag{3.55}
$$

and the feedback (3.45) simplifies to (3.25). Since $A_1 + L_b C$ is Hurwitz, the e_b-dynamics are exponentially stable. $(0, I)$ is a locally exponentially stable equilibrium of the (e, E) dynamics, as shown in Lemma 3.6. In other words, the point $(0, 0, I)$ is a locally exponentially stable equilibrium for (3.55), thus $(0, I) \times (0, 0, I)$ is a locally exponentially stable equilibrium of the dynamics (3.46) restricted to the set where V_4 is zero. The Theorem by Iggidr and Sallet [62, Theorem 5] then implies uniform stability of $(0, I) \times (0, 0, I)$ for the closed loop. To show the attractivity of $(0, I) \times (0, 0, I)$, we first show that $(0, I) \times (0, 0, I)$ is isolated in $\mathcal{M}_2 \times \mathcal{M}_3$. First consider the e_w-E_Θ-dynamics. The equilibria of the e_w-E_Θ-dynamics are given by $\{(e_w, E) \in \mathbb{R}^m \times SO(n) | e_w = 0, E_\Theta = E_\Theta^{\mathsf{T}}\}$ and are exactly the critical points of the function $(b, \Theta) \mapsto b^{\mathsf{T}} b + n - \mathrm{tr}(\Theta)$ (cf. Lemma 3.24a)). Since $(0, I)$ is a unique minimum of this function, there is an open set which contains $(0, I)$ and no other equilibria. If $e_w = \hat{w} - w = 0$, then the dynamics of e_b given by

$$
\dot{e}_b = (A_1 + L_b C_1) e_b + (v(y_1, \hat{w}) - v(y_1, w))
\tag{3.56}
$$

show that 0 is an isolated equilibrium of the e_b-dynamics. Hence there is an open set $\mathcal{U}_1$ in $\mathcal{M}_2$ which contains $(0, 0, I)$ and no other equilibrium. Furthermore, the property that $(0, I)$ is a unique minimum of $(b, \Theta) \mapsto b^{\mathsf{T}} b + n - \mathrm{tr}(\Theta)$ implies that there is an open set $\mathcal{U}_2$ in $\mathcal{M}_3$ which contains $(0, I)$ and no other equilibrium. The set $\mathcal{U}_1 \times \mathcal{U}_2$ is open in $\mathcal{M}_2 \times \mathcal{M}_3$ and, because of the uniform stability of $(0, I) \times (0, 0, I)$, we are certain that there is a positively invariant open $\mathcal{U}$ which is a subset of $\mathcal{U}_1 \times \mathcal{U}_2$ and which contains $(0, I) \times (0, 0, I)$. Therefore $\mathcal{U}$ contains no other equilibria than $(0, I) \times (0, 0, I)$, which implies that $(0, I) \times (0, 0, I)$ is isolated. As consequence of Theorem 1.17, all solutions converge to one of the connected components of $\mathcal{F} \times (0, 0, I)$. Since $(0, I) \times (0, 0, I)$ is the only connected component in $\mathcal{U}$, the solutions necessarily converge to $(0, I) \times (0, 0, I)$. This implies that $(0, I) \times (0, 0, I)$ is attractive. Overall, $(0, I) \times (0, 0, I)$ is uniformly stable and attractive, hence asymptotically stable.

To show that the remaining connected components $\mathcal{E}_k = \mathcal{M}_1 \times \mathcal{F}_k \times (0, 0, I)$ are unstable, we utilize Lemma 3.6 and the property that the $\mathcal{F}_k$ are the equilibria of (3.26). Consider again the dynamics of (3.46) restricted to the set where V_4 is zero, i.e. (3.55). On the set where V_4 is zero, (3.45) simplifies to (3.25), i.e. the (e, E) dynamics are given by (3.26). Lemma 3.6 shows that all equilibria except $(0, I)$ are unstable. Hence these equilibria are also unstable with respect to the dynamics of (3.46). ∎

We finally compare the result of Theorem 3.12 to classical output regulation in the sense of Isidori and Byrnes [66], Isidori and Marconi [67].

Remark 3.13. *Theorem 3.12 shows that the errors* $(e, E, e_w, e_b, E_\Theta)$ *typically converge to* $(0, I) \times (0, 0, I)$. *More precisely,* $(0, I) \times (0, 0, I)$ *is uniformly stable and attractive for the* $(e, E, e_w, e_b, E_\Theta)$-*dynamics if we consider the* $(e, E, e_w, e_b, E_\Theta)$-*dynamics as time varying system with* w, b_1, Θ_1 *as bounded time varying input functions. This means that a solution* $t \mapsto (x_1(t; x_0), x_2(t; x_0), x_3(t; x_0))$ *typically converges to the solution* $t \mapsto x_1(t; x_0) \times (0, I) \times (0, 0, I)$. *Consequently, the asymptotic time-behavior of the closed loop solutions* (3.13), (3.14), (3.32), (3.45) *and* (3.11) *is typically given by solutions with initial conditions from the set* $\mathscr{S}$ *given by*

$$\mathscr{S} = \big\{ (w, b_1, \Theta_1, b_2, \Theta_2, \hat{w}, \hat{b}_1, \hat{\Theta}_1) \in \mathscr{M} \,|\, \hat{w} = w,\ b_1 = \hat{b}_1 = b_2,\ \Theta_1 = \hat{\Theta}_1 = \Theta_2 \big\}. \quad (3.57)$$

$\mathscr{S}$ *is an invariant set for the closed loop. Therefore, the dynamics of the closed loop restricted to* $\mathscr{S}$ *can be considered as the steady state behavior for the closed loop in the spirit of Isidori and Byrnes [66].*

Theorem 3.12 implies convergence of the closed loop system despite the presence of multiple connected components of equilibria. Moreover, feedback and observer were designed independently. Note that the proposed feedback and observer are only one possible realization. More specifically, assume that a controller of the form (3.18) consists of an typically convergent observer for the exosystem (3.13) and (3.14) and a feedback that achieves typical convergence in the state feedback case. If this state feedback is such that V_3 from Lemma 3.11 is a height function on $\mathscr{N}$, then a certainty equivalence implementation of observer and state feedback solves 3.4, i.e. achieves typical convergence to the desired invariant set in the closed

.

3.2.5 Example

We consider an application example motivated by the Serret-Frenet equations. As mentioned in the introduction, the Serret-Frenet equations play an important role in robotics and computer graphics, see e.g. Agoston [3], Siciliano and Khatib [134]. For details on the Serret-Frenet equations from a control perspective, see Jurdjevic [71].

Let $\beta_1 : \mathbb{R} \to \mathbb{R}^3$ be a curve and assume that the curve is a solution of the system

$$\begin{aligned} \dot{\beta}_1 &= v(\kappa) H \beta_1 \\ \dot{R}_1 &= R_1 A(\kappa) \\ y_1 &= (c_1^\mathsf{T} \beta_1, R_1^\mathsf{T}), \end{aligned} \quad (3.58)$$

where

$$A(\kappa) = \begin{pmatrix} 0 & -\kappa_1 & 0 \\ \kappa_1 & 0 & -\kappa_2 \\ 0 & \kappa_2 & 0 \end{pmatrix}, \quad (3.59)$$

$(\beta_1, R_1) \in SE(3)$, $H \in \mathbb{R}^{3 \times 3}$ is stable, (c_1, H) is observable and $v : \mathbb{R}^2 \to \mathbb{R}$ is positive and bounded, i.e. there are constants $\delta_1 > 0$, $\delta_2 > \delta_1$ such that $\delta_1 < v(\kappa) < \delta_2$ for all $\kappa \in \mathbb{R}^2$. To recover the classical Serret-Frenet equations, one has to replace the right hand side of $\dot{\beta}_1$ in (3.58) by $R_1 e_1$, where $e_1 = (1, 0, 0)^\mathsf{T}$. In the Serret-Frenet case, the parameters $\kappa = (\kappa_1, \kappa_2)^\mathsf{T}$ are the curvatures of the curve β_1. We consider the modification (3.58), since it is in general not possible to infer the forward boundedness of $\beta_1(t)$ in case $\dot{\beta}_1 = R_1 e_1$. The assumptions on

H and v imply that $\beta_1(t)$ remains bounded, and that system (3.58) generates curves where the associated velocity vector has a bounded norm.

To relate system (3.58) to system (3.13), we consider the change of coordinates $\Theta_1 = R_1^{-1} = R_1^{\mathsf{T}}$ and $b_1 = R_1^{\mathsf{T}} \beta_1 = \Theta_1 \beta_1$. In these coordinates, system (3.58) is given by

$$
\begin{aligned}
\dot{\Theta}_1 &= \dot{R}_1^{\mathsf{T}} = -A(\kappa)R_1^{\mathsf{T}} = -A(\kappa)\Theta_1 \\
\dot{b}_1 &= \dot{R}_1^{\mathsf{T}}\beta_1 + R_1^{\mathsf{T}}\dot{\beta}_1 = \dot{R}_1^{\mathsf{T}}R_1 R_1^{\mathsf{T}}\beta_1 + v(\kappa)R_1^{\mathsf{T}}H\beta_1 \\
&= -A(\kappa)b_1 + v(\kappa)R_1^{\mathsf{T}}HR_1 b_1 \\
&= -A(\kappa)b_1 + v(\kappa)\Theta_1 H\Theta_1^{\mathsf{T}} b_1 \\
y &= (c_1^{\mathsf{T}}\Theta_1^{\mathsf{T}}b_1, \Theta_1).
\end{aligned}
\tag{3.60}
$$

Moreover, we assume that the time evolution of κ is determined by

$$
\begin{aligned}
\dot{w} &= Sw \\
\kappa &= Gw,
\end{aligned}
\tag{3.61}
$$

where $S \in \mathbb{R}^{m \times m}$ is skew-symmetric and $G \in \mathbb{R}^{2 \times m}$.

Similar to Section 3.2.1, we now consider a family of tracking problems in an output regulation setup, where the exosystem is given by (3.60) and (3.61). This exosystem differs from (3.13) and (3.14), because A_1 from (3.13) is replaced by $A(\kappa)$ in (3.60), which depends on κ. Furthermore, the skew symmetric matrix $Q(w)$ is replaced by the skew symmetric $-A(\kappa)$, where $\kappa = Gw$. The control system is given by

$$
\begin{aligned}
\dot{b}_2 &= -A(\kappa)b_2 + u \\
\dot{\Theta}_2 &= (-A(\kappa) + U)\Theta_2 \\
y_2 &= (\Theta_2 b_2, \Theta_2)
\end{aligned}
\tag{3.62}
$$

and corresponds to (3.11) in Section 3.2.1. The regulated output is

$$
(e, E) = (b_2 - b_1, \Theta_2^{-1}\Theta_1).
\tag{3.63}
$$

The goal of the output regulation problem is to achieve $(e(t), E(t)) \to (0, I)$ for $t \to \infty$ such that all states of the closed loop remain bounded.

In the following we show that the results of Section 3.2.1 can be applied to the problem consisting of (3.60), (3.61) and (3.62) despite the mentioned differences, which illustrates the flexibility of the proposed approach. We first discuss the observer design for (3.60) and (3.61). Second, we design an appropriate state feedback. Finally, we present some simulations of the closed loop utilizing the observer and a certainty equivalence implementation of the feedback.

First, we design the observer for the w and Θ_1 states independently of the observer for b_1 states. The observer structure is motivated by the Luenberger observer, i.e.

$$
\begin{aligned}
\dot{\hat{w}} &= S\hat{w} + L_w(\hat{\Theta}_1, \Theta_1) \\
\dot{\hat{b}}_1 &= f_b(\hat{\kappa}, \hat{b}_1, \hat{\Theta}_1, y_1) \\
\dot{\hat{\Theta}}_1 &= \left(-A(\hat{\kappa}) + L_\Theta(\hat{\Theta}_1, \Theta_1)\right)\hat{\Theta}_1 \\
\hat{\kappa} &= G\hat{w} \\
\hat{y}_1 &= (c_1^{\mathsf{T}}\hat{\Theta}_1^{\mathsf{T}}\hat{b}_1, \hat{\Theta}_1),
\end{aligned}
\tag{3.64}
$$

75

where the vector field for $\hat{b}_1$ is discussed later. Define the observer errors by $e_w = \hat{w} - w$, $e_b = \hat{b}_1 - b_1$, $E_\Theta = \hat{\Theta}_1^{-1}\Theta_1$ and $e_\kappa = \hat{\kappa} - \kappa = Ge_w$. The observer error dynamics for e_w and E_Θ are given by

$$\dot{e}_w = Se_w + L_w(\hat{\Theta}_1, \Theta_1)$$
$$\dot{E}_\Theta = -A(e_\kappa)E_\Theta + [E_\Theta, A(\kappa)] + L_\Theta(\hat{\Theta}_1, \Theta_1)E_\Theta. \tag{3.65}$$

Now choose

$$(L_w(\hat{\Theta}_1, \Theta_1))_j = -G_{1j}((E_\Theta)_{12} - (E_\Theta)_{21})$$
$$- G_{2j}((E_\Theta)_{23} - (E_\Theta)_{32}) \tag{3.66}$$
$$L_\Theta(\hat{\Theta}_1, \Theta_1) = E_\Theta^\mathsf{T} - E_\Theta$$

and consider $V_5(e_w, E_\Theta) = \frac{1}{2}e_w^\mathsf{T}e_w + n - \mathrm{tr}(E_\Theta)$. With L_w and L_Θ as in (3.66), the e_w-dynamics and the E_Θ-dynamics are independent of the b_1-dynamics. Thus, V_5 is a suitable Lyapunov function candidate for the e_w-E_Θ-dynamics. Observe that the derivative $\dot{V}_5$ of V_5 in the direction of the e_w and the E_Θ vector field is given by

$$\dot{V}_5 = e_w^\mathsf{T}Se_w + e_w^\mathsf{T}L_w(\hat{\Theta}_1, \Theta_1)$$
$$- \mathrm{tr}\left(-A(e_\kappa)E_\Theta + L_\Theta(\hat{\Theta}_1, \Theta_1)E_\Theta\right). \tag{3.67}$$

Since $S \in \mathbb{R}^{m \times m}$ is skew-symmetric, $e_w^\mathsf{T}Se_w = 0$. Furthermore, L_w has been chosen such that $e_w^\mathsf{T}L_w(\hat{\Theta}_1, \Theta_1) + \mathrm{tr}(A(e_\kappa)E_\Theta) = 0$. Hence $\dot{V}_5 = -\mathrm{tr}\left((E_\Theta^\mathsf{T} - E_\Theta)E_\Theta\right)$ which implies convergence towards $\{E_\Theta \in SO(n) | E_\Theta = E_\Theta^\mathsf{T}\}$. The same arguments as in Lemma 3.6 (see Section 3.2.2) imply that I is the only asymptotically stable equilibrium for E_Θ and that E_Θ converges typically towards I in the sense of Definition 3.1. We now show briefly that $E_\Theta - I$ is asymptotically e_κ-detectable. Similar to the proof of Lemma 3.9, we drop in the explicit dependence on the initial conditions for the solution components to show the detectability. More precisely, we write $t \mapsto e_\kappa(t)$ for the e_κ-component of the solution of (3.34), and likewise for the e_w and the E_Θ-component. Because of (3.67), the e_w-component of the error dynamics remains bounded and therefore $e_\kappa = Ge_w$ remains bounded. The E_Θ-component ist bounded since $E_\Theta \in SO(n)$. With the same arguments as in the proof of Lemma 3.9, we conclude $\dot{E}_\Theta(t) \to 0$ for $t \to \infty$. Since $L_\Theta(\hat{\Theta}_1, \Theta_1) = 0$ if $E_\Theta = I$, $\dot{E}_\Theta(t) \to 0$ for $t \to \infty$ in (3.65) implies $A(e_\kappa(t)) = A(Ge_w(t)) \to 0$ for $t \to \infty$. Hence, due to the structure of $A(e_\kappa)$, $e_\kappa(t) = Ge_w(t) \to 0$ for $t \to \infty$.

Next, we design the observer for b_1. Denote the observer state of b_1 by $\hat{b}_1$ and the error by $e_b = \hat{b}_1 - b_1$. Since we can measure Θ_1, we know $e_b = \hat{b}_1 - b_1 = \hat{\Theta}_1\hat{\beta}_1 - \Theta_1\beta_1 = \Theta_1(\hat{\beta}_1 - \beta_1)$, i.e. with $e_\beta = \hat{\beta}_1 - \beta_1$, we have $e_b = \Theta_1 e_\beta$. Consequently, we can design the observer in β_1-coordinates, which is notably simpler than in the b_1-coordinates. Consider

$$\dot{\beta}_1 = v(\kappa)H\beta_1$$
$$y_1 = (c_1^\mathsf{T}\beta_1, \Theta_1) \tag{3.68}$$

and let the observer dynamics be given by

$$\dot{\hat{\beta}}_1 = v(\hat{\kappa})H\hat{\beta}_1 + L_\beta(\hat{\kappa}, \hat{\beta}_1, y_1)$$
$$\hat{y}_1 = (c_1^\mathsf{T}\hat{\beta}_1, \hat{\Theta}_1). \tag{3.69}$$

The dynamics for $e_\beta = \hat{\beta}_1 - \beta_1$ are given by

$$\dot{e}_\beta = v(\hat{\kappa})H\hat{\beta}_1 - v(\kappa)H\beta_1 + L_\beta(\hat{\kappa}, \hat{\beta}_1, y_1)$$
$$= v(\hat{\kappa})He_\beta + (v(\hat{\kappa}) - v(\kappa))H\beta_1 + L_\beta(\hat{\kappa}, \hat{\beta}_1, y_1). \tag{3.70}$$

We choose

$$L_\beta(\hat{\kappa}, \hat{\beta}_1, y_1) = v(\hat{\kappa})\overline{L}_\beta(c_1^\mathsf{T}\hat{\beta}_1 - c_1^\mathsf{T}\beta_1) = v(\hat{\kappa})\overline{L}_\beta c_1^\mathsf{T} e_\beta, \tag{3.71}$$

where $\overline{L}_\beta \in \mathbb{R}^3$ is a constant gain. Utilizing (3.71) implies

$$\dot{e}_\beta = v(\hat{\kappa})(H + \overline{L}_\beta c_1^\mathsf{T})e_\beta + (v(\hat{\kappa}) - v(\kappa))H\beta_1. \tag{3.72}$$

Since (c_1, H) is observable, we can place the poles of $H + \overline{L}_\beta c_1^\mathsf{T}$ in the left half of the complex plane, and since $v(\hat{\kappa})$ is a positive bounded function, this implies exponential stability of $\dot{e}_\beta = v(\hat{\kappa})(H + \overline{L}_\beta c_1^\mathsf{T})e_\beta$. Since the κ and the β_1-component remain bounded and $(v(\hat{\kappa}(t)) - v(\kappa(t)))H\beta_1(t) \to 0$ for $t \to 0$, we get $e_\beta(t) \to 0$ for $t \to \infty$. Furthermore, the previous discussion implies that the e_β-component remains bounded. As a consequence, the vector field f_b in (3.64) is given by the transformation of the vector field for $\hat{\beta}_1$ in (3.69) into $\hat{b}_1$-coordinates.

Finally we design the state feedback. Consider the dynamics of the regulated output, i.e.

$$\begin{aligned} \dot{e} &= -A(\kappa)e - v(\kappa)\Theta_1 H\Theta_1^\mathsf{T}b_1 + u \\ \dot{E} &= (-A(\kappa) + EA(\kappa)E^\mathsf{T} + U)E. \end{aligned} \tag{3.73}$$

If we choose the feedback as

$$\begin{aligned} u &= A(\kappa)e + v(\kappa)\Theta_1 H\Theta_1^\mathsf{T}b_1 - e \\ U &= A(\kappa) - EA(\kappa)E^\mathsf{T} + E^\mathsf{T} - E, \end{aligned} \tag{3.74}$$

then the error dynamics (3.73) are the same as (3.26). Consequently, Lemma 3.6 and Lemma 3.7 imply that the state feedback (3.74) achieves asymptotic stability of $(0, I)$ for almost all initial conditions. Utilizing the certainty equivalence feedback

$$\begin{aligned} u &= A(\hat{\kappa})\hat{e} + v(\hat{\kappa})\Theta_1 H\Theta_1^\mathsf{T}\hat{b}_1 - \hat{e} \\ U &= A(\hat{\kappa}) - EA(\hat{\kappa})E^\mathsf{T} + E^\mathsf{T} - E, \end{aligned} \tag{3.75}$$

where $\hat{e} = b_2 - \hat{b}_1$, guarantees that the closed loop solutions consisting of the components w, b_1, Θ_1, $\hat{w}$, $\hat{b}_1$, $\hat{\Theta}_1$, b_2 and Θ_2 remain bounded. More precisely, the $\Theta_1, \hat{\Theta}_1, \Theta_2$ components are bounded, since $\Theta_1, \hat{\Theta}_1, \Theta_2 \in SO(n)$ and $SO(n)$ is compact. The w-component of the closed loop solution is bounded since $S = -S^\mathsf{T}$. The b_1-component remains bounded since β_1 remains bounded and $b_1 = \Theta_1\beta_1$. The boundedness of the $\hat{w}$-component is guaranteed by the boundedness of the observer error $e_w = \hat{w} - w$. The $\hat{b}_1$-component remains bounded since e_β remains bounded which implies that $e_b = \hat{b}_1 - b_1 = \Theta_1 e_\beta$ remains bounded. The boundedness of the b_2-component follows with the same argument as given in the proof of Theorem 3.12. Since the observer (3.64) is convergent and the solutions remain bounded, Theorem 3.12 implies that the closed loop consisting of the exosystem (3.60) and (3.61), the system (3.62) and the observer (3.64) achieves convergence of the regulated output $(e(t), E(t)) \xrightarrow{\text{typ}} (0, I)$ for $t \to \infty$.

In the following, we show exemplary solutions of the closed loop. The parameters chosen for the simulation are as follows:

$$S = \begin{pmatrix} 0 & 0.4 & 0 & 0 & 0 & 0 & 0 \\ -0.4 & 0 & 0 & 0 & 0 & 0 & 0 \\ 0 & 0 & 0 & 0.05 & 0 & 0 & 0 \\ 0 & 0 & -0.05 & 0 & 0 & 0 & 0 \\ 0 & 0 & 0 & 0 & 0 & 0.025 & 0 \\ 0 & 0 & 0 & 0 & -0.025 & 0 & 0 \\ 0 & 0 & 0 & 0 & 0 & 0 & 0 \end{pmatrix}$$

$$G = \begin{pmatrix} 0.1 & 0 & 0 & 0.2 & 0 & 0.02 & 0.15 \\ 0 & 0.2 & 0 & 0 & 0 & 0 & 0.91 \end{pmatrix} \qquad (3.76)$$

$$w(0) = \begin{pmatrix} 1 & 0 & 1 & 0 & 1 & 0 & 1 \end{pmatrix}^{\mathsf{T}}$$

$$b_1(0) = \begin{pmatrix} 0.1 & 0.1 & 0.1 \end{pmatrix}^{\mathsf{T}}$$

$$\Theta_1(0) = \begin{pmatrix} 0.37 & 0.77 & -0.51 \\ -0.90 & 0.43 & 0.00 \\ 0.220 & 0.46 & 0.86 \end{pmatrix}$$

$$\hat{w}(0) = 0, \ \hat{b}_1(0) = 0, \ \hat{\Theta}_1(0) = I.$$

The plots in Figure 3.3 show a typical time evolution of e_b in the left column and a typical time evolution of e in the right column. Figure 3.4 shows a typical time evolution of E_{Θ}. Figure 3.5 shows a typical time evolution of E.

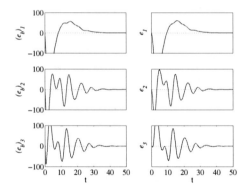

Figure 3.3: Exemplary time evolution of e_b in the left column and of e in the right column.

3.2.6 Summary

In this section we propose a solution for a class of output regulation problems evolving on the special Euclidean group $SE(n)$. The solution is given in the natural representation of $SE(n)$ and the solution is global. We showed that it is possible to render the desired equilibrium of the considered output regulation problem the only asymptotically stable equilibrium despite the

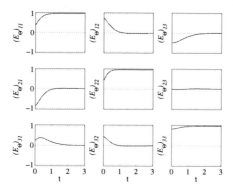

Figure 3.4: Exemplary time evolution of E_Θ.

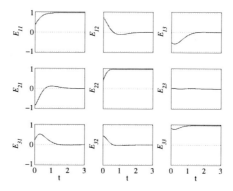

Figure 3.5: Exemplary time evolution of E.

presence of additional equilibria. The exponential stability of the desired equilibrium shows that the approach includes a certain measure of local robustness. The proposed controller consists of a certainty equivalence implementation of a state feedback and an observer. The state feedback can be designed independently of the observer and the results in this section can be interpreted as a nonlinear separation principle for the considered class of control problems.

3.3 Global output regulation for rigid body systems

In this Section, we are going to present a solution to a global nonlinear output regulation problem for the rotational dynamics of rigid bodies. The class of problems we consider includes attitude tracking problems for rigid bodies. The contributions of this Section are summarized in the following. First, we show how to extend a global controller design method for output regulation problems on $SE(n)$ from Section 3.2 to output regulation problems for the rotational rigid body dynamics. The model class we consider here includes multibody systems with one controlled rigid body. Similar to the design in Section 3.2, we present a two step design which consists of a state feedback and an observer which are designed independently. In the rigid body control problem, one does not have direct control of the attitudes. Therefore, the state feedback design is more complicated than for the $SE(n)$ case. We resolve this by employing a backstepping-like technique. The observer design also differs due to the additional dynamics. More precisely, we include an LMI-based design step into the observer design. We show that the states of the closed loop system remain bounded and that the proposed method achieves typical convergence of the error in the sense of Definition 3.1. Together with the independent state feedback design this establishes a separation principle for rigid body control problems. Furthermore, we demonstrate the applicability of the theory for a realistic application scenario, i.e. we apply the method to the attitude tracking problem of a satellite consisting of three rigid bodies. The task of the controller is to achieve attitude tracking for one of the three bodies despite uncontrolled but bounded motions of the attached bodies and despite presence of harmonic disturbances which cannot be measured. This section is based on Schmidt et al. [124].

The remainder of the Section is structured as explained in the following. We give a detailed problem statement in Section 3.3.1 and a short literature review in Section 3.3.2. In Section 3.3.3, we present our state feedback design for the considered problem class. We design the observer for the states that are not available from measurements in Section 3.3.4. In Section 3.3.5 we establish the stability of the closed loop which uses a certainty equivalence implementation of the state feedback. We present an application scenario in Section 3.3.6. Finally, we give a summary in Section 3.3.7.

3.3.1 Problem statement

The equations of motion for the rotational motion of the rigid body in an inertial frame are given by

$$\dot{\Theta} = Q(\Theta)\Theta$$
$$\widehat{J\omega} = M, \tag{3.77}$$

for more details on the equations of motion of a rigid body consider Murray et al. [102]. Additional details on the kinematic equations in the context of this thesis are given in Section A (Appendix). In the remainder of the Section we abbreviate the angular momentum $J\omega$ of a rigid body by z, i.e. we write (3.77) in the form

$$\dot{z} = M \quad \text{and} \quad \dot{\Theta} = Q\left(J^{-1}z\right)\Theta. \tag{3.78}$$

In the following, we explain the specific class of control systems and the specific class of exosystems utilized in this Section. The setup we utilize for the rigid body problem is similar to the setup we utilized for the problem on $SE(n)$ in Section 3.2. Therefore, we utilize the same control approach as in Section 3.2, i.e. a two-step observer based output feedback controller. However, due to differences in the dynamics of the control system and due to differences in the exosystem, the feedback design and the observer design differ from the feedback design and the observer design presented in Section 3.2. In other words, the ideas behind the output feedback design and the separate design steps remain similar, but the differences require additional effort to prove the desired properties for the closed loop. We highlight the differences in the following after introducing the control system and the exosystem. The complete setup is illustrated in Figure 3.6. The goal is that the attitude of the control system tracks a family of

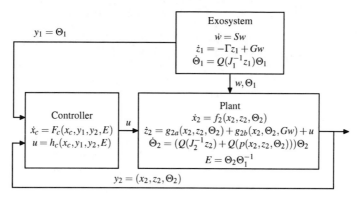

Figure 3.6: Illustration of the output regulation setup.

reference attitudes. The state space of the closed loop system is denoted by $\mathcal{M}$. Just as in the case of the problem on $SE(n)$ in Section 3.2, $\mathcal{M}$ consists of the state space of the exosystem $\mathcal{M}_1$, the state space of the control system $\mathcal{M}_2$ and the state space of the controller $\mathcal{M}_3$ (see also Section 3.3.4). Similar to the system description in Section 3.2.1, we use the subscript "1" for the states of the exosystem and the subscript "2" for the states of the control system. This is consistent with the subscripts of the corresponding state spaces. The class of control systems that we consider in this Section has the form

$$
\begin{aligned}
\dot{x}_2 &= f_2(x_2, z_2, \Theta_2) \\
\dot{z}_2 &= g_{2a}(x_2, z_2, \Theta_2) + g_{2b}(x_2, \Theta_2, Gw) + u \\
\dot{\Theta}_2 &= (Q\left(J_2^{-1} z_2\right) + Q\left(p(x_2, z_2, \Theta_2)\right))\Theta_2 \\
y_2 &= (x_2, z_2, \Theta_2),
\end{aligned}
\tag{3.79}
$$

where $x_2 \in \mathbb{R}^l$, $z_2 \in \mathbb{R}^3$, $\Theta_2 \in SO(3)$, $w \in \mathbb{R}^m$, $G \in \mathbb{R}^{3 \times m}$ and u denote the control input. y_2 denotes the output of the system. In attitude control problems and attitude estimation problems it is common that the attitude is available from measurements, see e.g. Wen and Kreutz-Delgado [147], Chaturvedi et al. [26]. The state space of (3.79) is denoted by $\mathcal{M}_2$, i.e.

$$
\mathcal{M}_2 = \mathbb{R}^l \times \mathbb{R}^3 \times SO(3).
\tag{3.80}
$$

We assume that $p : \mathcal{M}_2 \to \mathbb{R}^3$, $f_2 : \mathcal{M}_2 \to \mathbb{R}^l$, $g_{2a} : \mathcal{M}_2 \to \mathbb{R}^3$ and $g_{2b} : \mathbb{R}^3 \times SO(3) \times \mathbb{R}^3 \to \mathbb{R}^3$ are smooth. The difference between the control system (3.79) considered here and the control system (3.11) considered in Section 3.2 are the additional x_2-dynamics, the z_2-dynamics and the way the control input can affect the system. The x_2-dynamics account for additional uncontrolled dynamics, in the multibody satellite application scenario in Section 3.3.6 these are the dynamics of the uncontrolled bodies. In contrast to (3.11), where we could control the b_2- and the Θ_2-dynamics, the only controlled part in (3.79) are the z_2-dynamics. However, we still want to solve an attitude control problem, which means that state we eventually want to control is Θ_2. This requires a different strategy for the feedback design and consequently additional steps in the proof for the convergence of the closed loop. In Section 3.3.3, we present

81

a state feedback design based on a backstepping-like technique which resolves the additional difficulties.

A part of the exosystem is a system on $SO(3)$ similar to a part of the control system (3.79), since our considerations are motivated by tracking applications. More precisely, we assume that the exosystem is of the form

$$\begin{aligned}
\dot{z}_1 &= -\Gamma z_1 + Gw \\
\dot{\Theta}_1 &= Q\left(J_1^{-1}z_1\right)\Theta_1 \\
y_1 &= \Theta_1,
\end{aligned} \qquad (3.81)$$

where $z_1 \in \mathbb{R}^3$, $\Theta_1 \in SO(3)$, $-\Gamma \in \mathbb{R}^{3\times3}$ is stable (Hurwitz). The solutions of the system (3.81) are rigid body motions for a system affected by the external momentum Gw. We assume w to be a solution of the differential equation

$$\dot{w} = Sw, \qquad (3.82)$$

where $S \in \mathbb{R}^{m\times m}$, $S = -S^{\mathsf{T}}$ and S does not have any zero eigenvalue. This is a difference to the problem considered in Section 3.2.1 which we need to guarantee boundedness. More precisely, since $S = -S^{\mathsf{T}}$, all solutions of (3.82) are bounded. The fact that S does not have any zero eigenvalue implies that the solutions of (3.82) are sines and cosines. Since $-\Gamma$ is stable and the solutions of (3.82) are sines and cosines, the z_1-part of a solution of (3.81) remains bounded, also in the case $\Gamma = 0$. We denote the state space of the exosystem (3.82) and (3.81) by $\mathcal{M}_1$, i.e.

$$\mathcal{M}_1 = \mathbb{R}^m \times \mathbb{R}^3 \times SO(3). \qquad (3.83)$$

In the application scenario we consider in Section 3.3.6, the exosystem of the form (3.81), (3.82) represents a reference system. The difference between the exosystem system (3.81), (3.82) considered here and the exosystem (3.13), (3.14) considered in Section 3.2 are the z_1-dynamics. This requires an extension for the observer design and consequently additional steps in the proof of the convergence of the observer, both of which we present in Section 3.3.4.

The control goal is the regulation of the error between the attitude Θ_2 and the reference attitude Θ_1. We utilize a common way to express the error E between two attitudes, i.e.

$$E = \Theta_2\Theta_1^{-1}, \qquad (3.84)$$

see e.g. Wen and Kreutz-Delgado [147], Lageman et al. [77]. If $E(t) \to I$ for $t \to \infty$, then we have $\Theta_2(t) = \Theta_1(t)$ for $t \to \infty$, as desired in tracking problems. Note that the requirement in output regulation is that the closed loop system achieves asymptotic tracking of a family of solutions generated by the exosystem instead of tracking of a single known reference. The controller is assumed to be of the form

$$\begin{aligned}
\dot{x}_c &= F_c(x_c, y_1, y_2, E) \\
u &= h_c(x_c, y_1, y_2, E),
\end{aligned} \qquad (3.85)$$

where x_c is the state of the controller and F_c, h_c are assumed to be smooth. Note that we formally list E in the arguments of the vector field and in the output function (3.85), ignoring the redundancy due to the arguments Θ_1, Θ_2. With the given system descriptions we summarize the output regulation problem we want to solve as follows:

Problem statement 3.14. *Find a controller of the form* (3.85) *such that the closed loop consisting of* (3.79), (3.81), (3.82) *and* (3.85) *has the following properties:*

a) . The closed loop achieves

$$\lim_{t\to\infty} E(t) \xrightarrow{typ} I \tag{3.86}$$

in the sense of Definition 3.1.

b) The zero error manifold Δ, *given by*

$$\Delta = \{(\Theta_1, \Theta_2) | \Theta_2 \Theta_1^{-1} = I\}, \tag{3.87}$$

is positively invariant for the closed loop system.

c) The solutions of the closed loop system remain bounded for all initial conditions.

We ask for typical convergence because of the nontrivial geometry of $SO(3)$. As explained in detail in Section 3.2.1, the existence of an asymptotically stable equilibrium for a vector field on $SO(3)$ implies the existence of additional equilibria. As a consequence of the presence of several equilibria, we ask for typical convergence of the error in the same sense as in Problem 3.4.

3.3.2 Preliminaries

In the following, we give a brief overview of relevant literature for the considered problem and the chosen solution approach. The attitude control and attitude tracking problem are important examples of nonlinear control problems and received considerable attention since many decades, see e.g. Meyer [95], Crouch [33] or Wen and Kreutz-Delgado [147]. Despite being a long treated subject, the field is still very active an offers challenging problems Chaturvedi et al. [26].

One important aspect in attitude control is the choice of attitude representation, i.e. in which mathematical way to describe the attitude of the rigid body. As explained before, we utilize the natural way to describe the attitude, i.e. the attitude is described by a rotation matrix $\Theta \in SO(3)$. For an overview of the attitude representation, see Shuster [133]. An alternative representation of the attitude are for example the Euler angles, which are coordinates on the three dimensional manifold $SO(3)$, see e.g. [49, Section 4.4]. Another representation are the unit quaternions, which is the set S^3 with a specific group structure, for details see e.g. Chaturvedi et al. [26]. The main point for our investigations is that none of the alternative representations is homeomorphic to $SO(3)$. As a consequence, a controller which was designed in an alternative representation does not imply the same stability properties of the equilibria and convergence properties for the controlled system on $SO(3)$. For more specific details of the advantages of the natural representation, see e.g. Bhat and Bernstein [15] or Chaturvedi et al. [26].

A specific choice for the problem setup given in Section 3.3.1 is the output regulation framework. The utilization of output regulation for rigid body control is not new, for a semi-global internal model based approach for various applications utilizing the unit quaternion representation see e.g. Isidori et al. [68]. Another semi-global approach which utilizes the unit quaternion representation, and where the problem description bears great resemblance to the one considered here, is given in Chen and Huang [27]. There, the authors also point out one of the major advantages of the output regulation framework in this context, which is the ability of disturbance rejection for harmonic disturbances. Another contribution where the problem setup is similar to ours is Sanyal et al. [116]. One significant difference of this contribution to the one presented here is the stability analysis. More precisely, the authors analyze the closed loop stability properties with the help of a single Lyapunov function. Even though this approach works well for the setup proposed in Sanyal et al. [116], it is less flexible than the two-step design method which we present here.

An especially motivating contribution to investigate the two-step design approach, i.e. the independent design of a feedback and an observer, was Salcudean [115]. This paper considers the design of an angular velocity observer. In Salcudean [115], the authors ask specifically for designs for rigid body control problems which allow the separate design of a feedback and an observer. To the best of the authors knowledge, the only contribution with a separation property for systems on $SO(3)$ aside from this thesis is Maithripala et al. [88]. The paper Maithripala et al. [88] however relies on strong convergence properties of the observer error, which might be difficult to guarantee for the considered problem class.

In the following sections we present the details of our approach to solve the problem described in Section 3.3.1.

3.3.3 State feedback design

First, we derive a necessary condition for the state feedback that guarantees the invariance of the zero error manifold Δ given in (3.87).

Lemma 3.15. *Assume a controller of the form* (3.85) *solves the regulation problem as stated in Problem statement 3.14. If* $\Theta_1 = \Theta_2$, *then the feedback u must have the form*

$$
u = -g_{2a}(x_2, z_2, \Theta_2) - g_{2b}(x_2, \Theta_2, Gw) - \left(J_2^{-1} + \frac{\partial p}{\partial z_2}(x_2, z_2, \Theta_2) \right)^{-1}
$$
$$
\left(-\widehat{J_2^{-1}} z_2 - \frac{\partial p}{\partial x_2}(x_2, z_2, \Theta_2)\dot{x}_2 - \frac{\partial p}{\partial \Theta_2}(x_2, z_2, \Theta_2)\dot{\Theta}_2 + \widehat{J_1^{-1}} z_1 \right). \tag{3.88}
$$

Proof. If a controller solves the problem in Problem statement 3.14, it renders Δ defined in (3.87) invariant, i.e. if $\Theta_2(0)\Theta_1^{-1}(0) = I$, then $E(t) = I$ for $t \geq 0$ and because of the smoothness of $t \mapsto E(t)$, this implies $\dot{E}(t) = 0$ and $\ddot{E}(t) = 0$ for $t \geq 0$. Then

$$
\dot{E} = \dot{\Theta}_2 \Theta_1^{-1} - \Theta_2 \Theta_1^{-1}\dot{\Theta}_1 \Theta_1^{-1}
$$
$$
= Q\left(J_2^{-1} z_2 \right) + Q\left(p(x_2, z_2, \Theta_2) \right) - Q\left(J_1^{-1} z_1 \right), \tag{3.89}
$$
$$
\ddot{E} = Q\left(\widehat{J_2^{-1}} z_2 + p(\widehat{x_2, z_2}, \Theta_2) - \widehat{J_1^{-1}} z_1 \right)
$$
$$
= Q\left(\widehat{J_2^{-1}} z_2 + \frac{\partial p}{\partial x_2}(x_2, z_2, \Theta_2)\dot{x}_2 \right)
$$
$$
+ Q\left(\frac{\partial p}{\partial \Theta_2}(x_2, z_2, \Theta_2)\dot{\Theta}_2 \right) + Q\left(-\widehat{J_1^{-1}} z_1 \right) \tag{3.90}
$$
$$
+ Q\left(\left(J_2^{-1} + \frac{\partial p}{\partial z_2}(x_2, z_2, \Theta_2) \right) (g_{2a}(x_2, z_2, \Theta_2) + g_{2b}(x_2, \Theta_2, Gw) + u) \right).
$$

The assertion is a consequence of $\ddot{E}(t) = 0$ for $t \geq 0$ and (3.90). ∎

Now we design a feedback under the assumption that we can measure the full state of the exosystem. The goal is to design a feedback u which depends on the states of the system

(3.79) and the exosystem (3.81), (3.82) and renders $E = I$ an almost globally asymptotically stable equilibrium for the E-dynamics. The state feedback control system written in error states is given by (3.81), (3.82) and

$$
\begin{aligned}
\dot{x}_2 &= f_2(x_2, z_2, E\Theta_1) \\
\dot{z}_2 &= g_{2a}(x_2, z_2, \Theta_2) + g_{2b}(x_2, \Theta_2, Gw) + u \\
\dot{E} &= \left(Q\left(J_2^{-1}z_2\right) + Q(p(x_2, z_2, E\Theta_1)) - EQ\left(J_1^{-1}z_1\right)E^{\mathsf{T}} \right) E.
\end{aligned}
\tag{3.91}
$$

We utilize a backstepping-like technique for the feedback design of the (z_2, E)-dynamics, see e.g. Krstić et al. [75]. The goal is to render I attractive for the E-dynamics. This is achieved for almost all initial conditions if $\dot{E} = (E^{\mathsf{T}} - E)E$, see Corollary 3.31. Associated to the desired E-dynamics is the control Lyapunov function $E \mapsto n - \operatorname{tr}(E)$. We get the desired error dynamics $\dot{E} = (E^{\mathsf{T}} - E)E$ if

$$
Q\left(J_2^{-1}z_2\right) = -Q(p(x_2, z_2, E\Theta_1)) + EQ\left(J_1^{-1}z_1\right)E^{\mathsf{T}} + E^{\mathsf{T}} - E.
\tag{3.92}
$$

Define $A : \mathbb{R}^3 \times SO(3) \times \mathcal{M}_2 \to T_I SO(3)$ by the right hand side of (3.92) and let $\tilde{z}_2$ be defined by

$$
Q(\tilde{z}_2) = Q(J_2^{-1}z_2) - A(z_1, \Theta_1, x_2, z_2, E).
\tag{3.93}
$$

Suppose that $T : \mathcal{M}_1 \times \mathcal{M}_2 \to \mathcal{M}_1 \times \mathcal{M}_2$ defined by

$$
T(w, z_1, \Theta_1, x_2, z_2, E) = (w, z_1, \Theta_1, x_2, \tilde{z}_2, E)
\tag{3.94}
$$

is a diffeomorphism that transforms the z_2-dynamics in (3.91) into $\tilde{z}_2$ coordinates, then we obtain

$$
\dot{\tilde{z}}_2 = \left(J_2^{-1} + \frac{\partial p}{\partial z_2}(x_2, z_2, E\Theta_1) \right) \dot{z}_2 + h(Gw, z_1, \Theta_1, x_2, \tilde{z}_2, E),
\tag{3.95}
$$

where

$$
\begin{aligned}
h(Gw, z_1, \Theta_1, x_2, \tilde{z}_2, E) &= \frac{\partial p}{\partial x_2}(x_2, z_2, E\Theta_1)\dot{x}_2 + \overbrace{J_2^{-1}z_2} \\
+ \frac{\partial p}{\partial \Theta_2}(x_2, z_2, E\Theta_1)(\dot{E}\Theta_1 + E\dot{\Theta}_1) &- Q^{-1}\left(EQ\overbrace{\left(J_1^{-1}z_1\right)}E^{\mathsf{T}} \right) - Q^{-1}\left(\overbrace{E^{\mathsf{T}} - E} \right),
\end{aligned}
\tag{3.96}
$$

where we have to substitute z_2 by the appropriate component of $T^{-1}(w, z_1, \Theta_1, x_2, \tilde{z}_2, E)$. In the following, we implicitly assume that this substitution is carried out if we write z_2 to keep the presentation short. Assume that $J_2^{-1} + \frac{\partial p_2(x_2, z_2, \Theta_2)}{\partial z_2}$ is invertible and if we choose u as

$$
\begin{aligned}
u(Gw, z_1, \Theta_1, x_2, \tilde{z}_2, E) &= -g_{2a}(x_2, z_2, \Theta_2) - g_{2b}(x_2, \Theta_2, Gw) \\
&- \left(J_2^{-1} + \frac{\partial p}{\partial z_2}(x_2, z_2, E\Theta_1) \right)^{-1} (k(\tilde{z}_2, E) - h(Gw, z_1, \Theta_1, x_2, \tilde{z}_2, E)),
\end{aligned}
\tag{3.97}
$$

then we immediately have:

Lemma 3.16. *Assume that the map* $T : \mathcal{M}_1 \times \mathcal{M}_2 \to \mathcal{M}_1 \times \mathcal{M}_2$ *defined by (3.94) is a diffeomorphism and that* $J_2^{-1} + \frac{\partial p}{\partial z_2}(x_2, z_2, E\Theta_1)$ *is everywhere non-singular. Moreover, suppose* u *is* *chosen as in (3.97). Then the closed loop dynamics are given by the exosystem (3.81), (3.82), the first equation in (3.91) and*

$$\dot{\tilde{z}}_2 = k(\tilde{z}_2, E)$$
$$\dot{E} = \left(Q(\tilde{z}_2) + E^\mathsf{T} - E \right) E. \tag{3.98}$$

In the next lemma we give a choice for $k(\tilde{z}_2, E)$ such that we have the desired stable error dynamics.

Lemma 3.17. *Consider (3.98) where* $k : \mathbb{R}^3 \times SO(3)$ *is*

$$k(\tilde{z}_2, E) = -\tilde{z}_2. \tag{3.99}$$

Then the following holds:

a) *The ω-limit set of any solution is contained in the set of equilibria*

$$\mathscr{F} = \{ (\tilde{z}_2, E) \in \mathbb{R}^3 \times SO(3) | \tilde{z}_2 = 0, E^\mathsf{T} = E \}. \tag{3.100}$$

b) *The equilibrium $(0, I)$ is locally exponentially stable and all other equilibria are unstable.*

c) *The set of initial conditions that converge towards $(0, I)$ is dense in $\mathbb{R}^3 \times SO(3)$ and the set of initial conditions that converge to the other equilibria has measure zero.*

Proof. a) The closed loop error dynamics with the feedback (3.99) are given by

$$\dot{\tilde{z}}_2 = -\tilde{z}_2$$
$$\dot{E} = \left(Q(\tilde{z}_2) + E^\mathsf{T} - E \right) E. \tag{3.101}$$

The equilibria are determined by setting the right hand side of (3.101) equal to zero. The vector field for $\tilde{z}_2$ then implies $\bar{\tilde{z}}_2 = 0$ for an equilibrium $\bar{\tilde{z}}_2$ which in turn implies $Q(\bar{\tilde{z}}_2) = 0$, since $Q(\cdot)$ is an isomorphism. Setting the right hand side of the vector field for E equal to zero and utilizing $Q(\bar{\tilde{z}}_2) = 0$ implies $2(\bar{E}^\mathsf{T} - \bar{E})\bar{E} = 0$ for an equilibrium $\bar{E}$. Since $\bar{E} \in SO(3)$ we thus need $\bar{E}^\mathsf{T} - \bar{E} = 0$. As a consequence, the equilibria of (3.101) are given by (3.100).

To determine the convergence to the set of equilibria, consider $V_1 : \mathbb{R}^3 \times SO(3) \to \mathbb{R}$ defined by

$$V_1(\tilde{z}_2, E) = \frac{1}{2}\tilde{z}_2^\mathsf{T} \tilde{z}_2 + 3 - \operatorname{tr}(E). \tag{3.102}$$

Then

$$\begin{aligned}
\dot{V}_1(\tilde{z}_2, E) &= -\tilde{z}_2^\mathsf{T} \tilde{z}_2 - \operatorname{tr}(Q(\tilde{z}_2)E) - \operatorname{tr}\left((E^\mathsf{T} - E)E \right) \\
&= -\tilde{z}_2^\mathsf{T} \tilde{z}_2 - \frac{1}{2}\operatorname{tr}\left((Q(\tilde{z}_2))^\mathsf{T} E^\mathsf{T} + Q(\tilde{z}_2)E \right) - \frac{1}{2}\operatorname{tr}\left(E^\mathsf{T}(E - E^\mathsf{T}) + (E^\mathsf{T} - E)E \right) \\
&= -\tilde{z}_2^\mathsf{T} \tilde{z}_2 - \frac{1}{2}\operatorname{tr}\left(Q(\tilde{z}_2)(E - E^\mathsf{T}) \right) - \frac{1}{2}\operatorname{tr}\left((E - E^\mathsf{T})(E^\mathsf{T} - E) \right) \\
&= -\tilde{z}_2^\mathsf{T} \tilde{z}_2 - \tilde{z}_2^\mathsf{T} Q^{-1}\left(E^\mathsf{T} - E \right) - \left(Q^{-1}\left(E^\mathsf{T} - E \right) \right)^\mathsf{T} Q^{-1}\left(E^\mathsf{T} - E \right),
\end{aligned} \tag{3.103}$$

where we utilized the identity $x^\mathrm{T} y = -\frac{1}{2}\operatorname{tr}(Q(x)Q(y))$ which holds for $x, y \in \mathbb{R}^3$. Thus, $\dot{V}_1$ is negative semidefinite and zero only on the set of equilibria. Therefore, the solutions of (3.98) with (3.99) converge to the set of equilibria given by (3.100).

b) To show the stability properties of the connected components of the equilibria we linearize the equations around the equilibria. Let (ξ, Ξ) be an element of $T_{(0,E_0)}(\mathbb{R}^3 \times SO(3))$, i.e. $\xi \in \mathbb{R}^3$ and $\Xi = UE_0$ where $U \in \mathbb{R}^{3\times 3}$ and $U = -U^\mathrm{T}$. Furthermore let $\gamma : (-\varepsilon, \varepsilon) \to \mathbb{R}^3$ and $\Gamma : (-\varepsilon, \varepsilon) \to SO(3)$ be differentiable curves such that $\gamma(0) = 0$, $\dot{\gamma}(0) = \xi$, $\Gamma(0) = E_0$, $\dot{\Gamma}(0) = \Xi = UE_0$. Then the differential of the right hand side of (3.101) is determined by

$$\frac{d}{dt}\Big|_{t=0} \left(\begin{array}{c} -\gamma(t) \\ (Q(\gamma(t)) + \Gamma(t)^\mathrm{T} - \Gamma(t))\Gamma(t) \end{array} \right). \tag{3.104}$$

Therefore the linearization of (3.101) is given by

$$\begin{aligned} \dot{\xi} &= -\xi \\ \dot{\Xi} &= Q(\xi)E_0 - E_0^\mathrm{T}\Xi - \Xi E_0. \end{aligned} \tag{3.105}$$

Thus, for $E_0 = I$, the linearization is

$$\begin{aligned} \dot{\xi} &= -\xi \\ \dot{\Xi} &= Q(\xi) - 2\Xi, \end{aligned} \tag{3.106}$$

which is asymptotically stable. In the other cases we have $E_0 = E_0^\mathrm{T} \neq I$. E_0 is orthonormally diagonalizable, i.e. $E_0 = \Pi^\mathrm{T}D\Pi$ for some diagonal D and orthonormal Π, where the columns of Π are eigenvectors of E_0. Since $E_0^\mathrm{T}E_0 = I$, we have $D^2 = I$ which implies that the eigenvalues of E_0 are ± 1. Since $E_0 \neq I$ and $\det E_0 = 1$, there is always an even number of negative eigenvalues. Let v_1, v_2 be the two eigenvectors with eigenvalue -1 and set $\overline{\Xi} = (v_1 v_2^\mathrm{T} - v_2 v_1^\mathrm{T})E_0 \in T_{E_0}SO(3)$. A calculation then shows

$$Q(\xi)E_0 - E_0^\mathrm{T}\overline{\Xi} - \overline{\Xi}E_0 = Q(\xi)E_0 + 2\overline{\Xi}E_0. \tag{3.107}$$

Consequently the linearization at $E_0 = E_0^\mathrm{T}$, $E_0 \neq I$ has an eigenvalue at 2 and a solution of the form

$$\begin{aligned} \xi(t) &= \exp(-It)\xi_0 \\ \Xi(t) &= \exp(2It)\overline{\Xi} + \int_0^t \exp(2I(t-\tau))Q(\xi(\tau))E_0 d\tau \end{aligned} \tag{3.108}$$

which shows that the equilibrium is unstable. Thus we have shown b).

c) We saw before that all solutions converge to the set of equilibria and that the set of equilibria consists of the two components $\mathcal{E}_0 = (0, I)$ and $\mathcal{E}_1 = \{0\} \times \{E \in SO(3) | \operatorname{tr}(E) = -1\}$. Lemma 3.24c) implies that $\mathcal{E}_1$ is connected and is a two-dimensional submanifold of $\mathbb{R}^3 \times SO(3)$. The linearization analysis in b) shows that the linearization around the elements of $\mathcal{E}_1$ has four eigenvalues with real part unequal to zero, hence $\mathcal{E}_1$ is normally hyperbolic. Furthermore, $\mathcal{E}_0$ has six eigenvalues with real part unequal to zero, thus it is normally hyperbolic as well. As a consequence, every solution converges to a single equilibrium, see e.g. [12, Proposition 4.1]. The convergence for almost all conditions follows with an argument similar to the one given in Lemma 3.29, see also [46, Proposition 1]. ∎

So far, our consideration of the state feedback design for (3.79) for the solution of the problem given in Problem statement 3.14 included only the (z_2, Θ_2)-dynamics and neglected the x_2-dynamics. As a consequence, the remaining issue for a state feedback solution of the Problem given in Problem statement 3.14 is the influence of the (Gw, z_2, Θ_2)-dynamics on the x_2-dynamics. We assume that the solutions of x_2 stay bounded under the influence of these inputs. This is guaranteed if the x_2-dynamics are input-to-state stable (ISS) with respect to Gw, z_2, Θ_2. For our application scenario, we show explicitly that this ISS property is guaranteed. Overall we have:

Corollary 3.18. *Let all assumptions of Lemma 3.16 hold. Furthermore, assume that the x_2-dynamics are input-to-state stable with respect to Gw, z_2, Θ_2. Then the state feedback law defined by (3.97) and (3.99) solves the problem given in Problem statement 3.14.*

3.3.4 Observer design

In this section we design an observer for the exosystem (3.81) and (3.82) that reconstructs (Gw, z_1, Θ_1) from Θ_1-measurements. Define the errors as $e_w = \hat{w} - w$, $e_z = \hat{z}_1 - z_1$ and $E_\Theta = \hat{\Theta}_1 \Theta_1^{-1}$ and utilize a Luenberger-type observer structure, i.e.

$$
\begin{aligned}
\dot{\hat{w}} &= S\hat{w} + l_w(y_1, \hat{w}, \hat{z}_1, \hat{\Theta}_1) \\
\dot{\hat{z}}_1 &= -\Gamma\hat{z}_1 + G\hat{w} + l_z(y_1, \hat{w}, \hat{z}_1, \hat{\Theta}_1) \\
\dot{\hat{\Theta}}_1 &= \left(Q\left(J_1^{-1}\hat{z}_1\right) + L(y_1, \hat{w}, \hat{z}_1, \hat{\Theta}_1)\right)\hat{\Theta}_1,
\end{aligned}
\tag{3.109}
$$

where l_w, l_z, L are smooth nonlinear observer gains. We denote the state space of (3.109) by

$$
\mathcal{M}_3 = \mathbb{R}^m \times \mathbb{R}^3 \times SO(3).
\tag{3.110}
$$

With the given errors, the error dynamics corresponding to (3.109) are

$$
\begin{aligned}
\dot{e}_w &= Se_w + l_w(y_1, e_w + w, e_z + z_1, E_\Theta \Theta_1) \\
\dot{e}_z &= -\Gamma e_z + Ge_w + l_z(y_1, e_w + w, e_z + z_1, E_\Theta \Theta_1) \\
\dot{E}_\Theta &= \dot{\hat{\Theta}}_1 \Theta_1^{-1} - \hat{\Theta}_1 \Theta_1^{-1} \dot{\Theta}_1 \Theta_1^{-1} \\
&= \left(Q\left(J_1^{-1}\hat{z}_1\right) + L(y_1, e_w + w, e_z + z_1, E_\Theta \Theta_1)\right) E_\Theta - E_\Theta Q\left(J_1^{-1}z_1\right) \\
&= \left(Q\left(J_1^{-1}e_z\right) + L(y_1, e_w + w, e_z + z_1, E_\Theta \Theta_1)\right) E_\Theta + [Q\left(J_1^{-1}z_1\right), E_\Theta],
\end{aligned}
\tag{3.111}
$$

where $[A, B] = AB - BA$ for $A, B \in \mathbb{R}^{n \times n}$. Since w and z_1 are not available from measurement, we need an assumption concerning the observability for our problem setup. More precisely, we have to stabilize the error dynamics of the observer utilizing the observer states and the error E_Θ. Therefore, the error dynamics need to have the property that, unless Ge_w and e_z did not converge to zero, E_Θ cannot converge to the identity either. In other words, we need that the output $y = E_\Theta - I$ is asymptotically (Ge_w, e_z)-detectable, see also Definition 3.8. The following lemma shows that this conditions holds in the current situation.

Lemma 3.19. *Consider the system given by (3.81) and (3.82) (exosystem) together with the observer error dynamics (3.111). Assume there are smooth observer gains l_w, l_z, L such that the solutions components $t \mapsto e_w(t; (x_0, \hat{x}_0))$, $t \mapsto e_z(t; (x_0, \hat{x}_0))$ of the system (3.81), (3.82) with (3.111) remain bounded for all initial conditions $x_0 = (w_0, z_{1,0}, \Theta_{1,0}) \in \mathcal{M}_1$ and $\hat{x}_0 = (e_{w_0}, e_{z_0}, E_{\Theta_0}) \in \mathcal{M}_3$. Furthermore, assume for the observer gains that $L(y_1, e_w + w, e_z + z_1, E_\Theta \Theta_1) = 0$ and $l_z(y_1, e_w + w, e_z + z_1, E_\Theta \Theta_1) = 0$ if $y_1 = \Theta_1$ and $E_\Theta = \Theta_1 \hat{\Theta}_1^{-1} = I$. Then the output $y = E_\Theta - I$ is asymptotically (Ge_w, e_z)-detectable.*

Proof. To avoid a too cluttered notation, we drop for this proof the explicit dependence of the solution components on the initial conditions. More precisely, we write $t \mapsto e_z(t)$ for the e_z-component of the solution of the system (3.81), (3.82) with (3.34), and likewise for the e_w and the E_Θ-component. Assume that $E_\Theta(t) \to I$ for $t \to \infty$. Since $E_\Theta(t) = E_\Theta(0) + \int_0^t \dot{E}_\Theta(\tau) d\tau$, we know that

$$\lim_{t \to \infty} \int_0^t \dot{E}_\Theta(\tau) d\tau = I - E_\Theta(0). \tag{3.112}$$

By assumption, the e_w and the e_z component of the solution of of the system (3.81), (3.82) with (3.111) remain bounded. The compactness of $SO(3)$ implies that the E_Θ-component of the solution is bounded. Differentiation of $t \mapsto \dot{E}_\Theta(t)$ reveals that $t \mapsto \ddot{E}_\Theta(t)$ remains bounded. Since $t \mapsto \ddot{E}_\Theta(t)$ remains bounded and is continuous, $t \mapsto \dot{E}_\Theta(t)$ is Lipschitz. This implies that $t \mapsto \dot{E}_\Theta(t)$ is uniformly continuous in t, see e.g. Amann and Escher [4, Chapter VI.2]. Since $t \mapsto \dot{E}_\Theta(t)$ is uniformly continuous and $\lim_{t \to \infty} \int_0^t \dot{E}_\Theta(\tau) d\tau$ exists and is finite, Barbalat's Lemma implies $\dot{E}_\Theta(t) \to 0$ for $t \to \infty$, see e.g. Krstić et al. [75, Lemma A.6]. Consequently, the right hand side of the third equation in (3.111) has to fulfill

$$0 = \lim_{t \to \infty} [Q\left(J_1^{-1}(t) z_1(t)\right), E_\Theta(t)] \tag{3.113}$$
$$\left(Q\left(J_1^{-1}(t) e_z(t)\right) + L(y_1(t), \hat{w}(t), \hat{z}_1(t), \hat{\Theta}_1(t))\right) E_\Theta(t).$$

Utilizing that $E_\Theta(t) \to I$ for $t \to \infty$ and that the observer gain L is smooth, we obtain that $Q\left(J_1^{-1}(t) e_z(t)\right) \to 0$ for $t \to \infty$. Since $J_1^{-1}(t)$ is non-singular for all t, this implies $e_z(t) \to 0$ for $t \to \infty$. As a result, we can continue the argument, i.e. we have $\int_0^t \dot{e}_z(\tau) d\tau \to -e_z(0)$ for $t \to \infty$, thus $\int_0^\infty \dot{e}_z(\tau) d\tau$ exists and is finite. We can again conclude that $t \mapsto \ddot{e}_z(t)$ remains bounded which implies the uniform continuity of $t \mapsto \dot{e}_z(t)$. Barbalat's Lemma implies then $\dot{e}_z(t) \to 0$ for $t \to \infty$. Therefore we necessarily have $Ge_w(t) \to 0$ for $t \to \infty$. ∎
The observer design is the result of the following:

Lemma 3.20. *Consider the error dynamics (3.111). Assume there are matrices $P_0 = P_0^\mathsf{T}$, $P_1 = P_1^\mathsf{T}$, P_2 such that*

$$\begin{pmatrix} P_0 & \frac{1}{2}P_2 \\ \frac{1}{2}P_2^\mathsf{T} & P_1 \end{pmatrix} > 0 \tag{3.114}$$

$$X + X^\mathsf{T} \leq 0, \tag{3.115}$$

where

$$X = \begin{pmatrix} P_0 S + \frac{1}{2}P_2 G & 0 \\ \frac{1}{2}(P_2^\mathsf{T} S - \Gamma^\mathsf{T} P_2^\mathsf{T}) + P_1 G & -P_1 \Gamma \end{pmatrix} \tag{3.116}$$

and let

$$l_w(E_\Theta) = -\frac{1}{2} P_0^{-1} P_2 l_z(E_\Theta) \tag{3.117}$$

$$l_z(E_\Theta) = (P_1 - \frac{1}{4} P_2^\mathsf{T} P_0^{-1} P_2)^{-1} (J_1^{-1})^\mathsf{T} \begin{pmatrix} E_{\Theta_{23}} - E_{\Theta_{32}} \\ E_{\Theta_{31}} - E_{\Theta_{13}} \\ E_{\Theta_{12}} - E_{\Theta_{21}} \end{pmatrix} \tag{3.118}$$

$$L(E_\Theta) = E_\Theta^\mathsf{T} - E_\Theta. \tag{3.119}$$

Then $(Ge_w(t), e_z(t), E_\Theta(t)) \xrightarrow{typ} (0, 0, I)$ for $t \to \infty$.

89

Proof. Consider $V_2 : \mathscr{M}_3 \to \mathbb{R}$ defined by

$$V_2(e_w, e_z, E_\Theta) = 3 - \operatorname{tr}(E_\Theta) + \frac{1}{2}[e_w^\mathsf{T} \; e_z^\mathsf{T}] \begin{pmatrix} P_0 & \frac{1}{2}P_2 \\ \frac{1}{2}P_2^\mathsf{T} & P_1 \end{pmatrix} \begin{pmatrix} e_w \\ e_z \end{pmatrix}$$

$$= 3 - \operatorname{tr}(E_\Theta) + \frac{1}{2}e_w^\mathsf{T} P_0 e_w + \frac{1}{2}e_z^\mathsf{T} P_1 e_z + \frac{1}{2}e_w^\mathsf{T} P_2 e_z. \tag{3.120}$$

Since $V_2(e_w, e_z, E_\Theta) > 0$ for $(e_w, e_z, E_\Theta) \neq (0, 0, I)$ and V_2 is radially unbounded with respect to e_w and e_z, the function V_2 is a Lyapunov function candidate to investigate the stability of $(0, 0, I)$. The derivative $\dot{V}_2$ of V_2 along the vector field (3.111) is

$$\begin{aligned}
\dot{V}_2 = & -\operatorname{tr}(L(E_\Theta)E_\Theta) \\
& - \operatorname{tr}\left(Q\left(J_1^{-1}e_z\right)E_\Theta\right) + e_z^\mathsf{T}\left(P_1 l_z(E_\Theta) + \frac{1}{2}P_2^\mathsf{T} l_w(E_\Theta)\right) \\
& - e_z^\mathsf{T} P_1 \Gamma e_z \\
& + e_w^\mathsf{T}(P_0 S + \frac{1}{2}P_2 G)e_w + e_w^\mathsf{T}(\frac{1}{2}S^\mathsf{T} P_2 + G^\mathsf{T} P_1^\mathsf{T} - P_2\Gamma)e_z \\
& + e_w^\mathsf{T}\left(\frac{1}{2}P_2 l_z(E_\Theta) + P_0 l_w(E_\Theta)\right).
\end{aligned} \tag{3.121}$$

In the following we discuss each expression in (3.121) separately. Because of (3.117) the last line in (3.121) is zero. Because of (3.115), the third and the fourth lines in (3.121) are equal to

$$\frac{1}{2}\left(e_w^\mathsf{T} \; e_z^\mathsf{T}\right)(X + X^\mathsf{T})\begin{pmatrix} e_w \\ e_z \end{pmatrix}. \tag{3.122}$$

The expansion of the term $\operatorname{tr}\left(Q\left(J_1^{-1}e_z\right)E_\Theta\right)$ explains the choice of l_z. More specifically,

$$\operatorname{tr}\left(Q\left(J_1^{-1}e_z\right)E_\Theta\right) = e_z^\mathsf{T}(J_1^{-1})^\mathsf{T}\begin{pmatrix} E_{\Theta_{23}} - E_{\Theta_{32}} \\ E_{\Theta_{31}} - E_{\Theta_{13}} \\ E_{\Theta_{12}} - E_{\Theta_{21}} \end{pmatrix}. \tag{3.123}$$

Therefore the second line in (3.121) gives:

$$\begin{aligned}
& - \operatorname{tr}\left(Q\left(J_1^{-1}e_z\right)E_\Theta\right) + e_z^\mathsf{T}\left(P_1 l_z(E_\Theta) + \frac{1}{2}P_2^\mathsf{T} l_w(E_\Theta)\right) \\
& = \operatorname{tr}\left(Q\left(J_1^{-1}e_z\right)E_\Theta\right) + e_z^\mathsf{T}(P_1 - \frac{1}{4}P_2^\mathsf{T} P_0^{-1} P_2)l_z(E_\Theta) = 0.
\end{aligned} \tag{3.124}$$

Finally, utilizing that $L(E_\Theta) = E_\Theta^\mathsf{T} - E_\Theta$ and (3.115), we obtain

$$\dot{V}_2 = \operatorname{tr}\left((E_\Theta - E_\Theta^\mathsf{T})E_\Theta\right) + \frac{1}{2}\left(e_w^\mathsf{T} \; e_z^\mathsf{T}\right)(X + X^\mathsf{T})\begin{pmatrix} e_w \\ e_z \end{pmatrix} \leq 0. \tag{3.125}$$

$\dot{V}_2$ is similar to $\dot{V}_1$ in the proof of Lemma 3.17, which implies the convergence $E_\Theta(t) \overset{typ}{\to} I$ for and the boundedness of e_w and e_z. Together with the (Ge_w, e_z)-detectability property from Lemma 3.19 this implies the typical convergence of (Ge_w, e_z, E_Θ) to $(0, 0, I)$. ∎

Remark 3.21. *The design involves the inequalities (3.114) and (3.115), which are linear matrix inequalities (LMIs) for P_0, P_1, P_2. For a given problem with data S, Γ, G a solution of (3.114) and (3.115) can be computed efficiently, see e.g. Boyd et al. [20]. To influence the convergence speed of E_Θ it is possible to choose $L(E_\Theta) = k(E_\Theta^\mathsf{T} - E_\Theta)$ with $k > 0$ instead of (3.119). Larger k would translate into faster convergence of E_Θ. However, our analysis does not indicate how this influences the convergence speed of e_w and e_z.*

3.3.5 Output feedback design

In this section we consider a controller consisting of the observer from Section 3.3.4 and the state feedback from Section 3.3.3 utilizing the estimated states of the observer. We show that the resulting controller provides a solution to the Problem given in Problem statement 3.14. Similar to Section 3.2, the proposed controller is a *certainty equivalence controller*, see e.g. Praly and Arcak [112]. Therefore, the term output feedback refers to the feedback which depends on the outputs of the exosystem, the estimates of the exosystem states and the full state of the control system. If only a part of the states of the control system is available from measurements and a convergent observer can be designed, then the results provided in this chapter can be adapted to obtain an analogous result which does not rely on the full state measurements of the control system. Here, the certainty equivalence controller consists of (3.109) and the feedback (3.97), where we replace Gw by the estimate $G\hat{w}$ and z_1 by the estimate $\hat{z}_1$, i.e.

$$u(G\hat{w}, \hat{z}_1, \Theta_1, x_2, z_2, E) - -g_{2a}(x_2, z_2, \Theta_2) \quad g_{2b}(x_2, \Theta_2, G\hat{w}) - \left(J_2^{-1} + \frac{\partial p}{\partial z_2}(x_2, z_2, \Theta_2)\right)^{-1}$$
$$\left(-J_2^{-1}z_2 + Q^{-1}\left(A(\hat{z}_1, \Theta_1, x_2, z_2, E)\right) - h(G\hat{w}, \hat{z}_1, \Theta_1, x_2, z_2, E)\right), \quad (3.126)$$

where A is defined after equation (3.92) and we utilize $\Theta_2 = E\Theta_1$. Hence, the closed loop consists of system (3.79), the exosystem (3.81) and (3.82), the observer (3.109), and the feedback (3.126). To analyze the closed loop behavior, we consider the closed loop in error states. The closed loop vector field in error states consists of the exosystem (3.81), (3.82), the first equation in (3.79) and

$$\dot{\tilde{z}}_2 = \left(J_2^{-1} + \frac{\partial p_2}{\partial z_2}\right)(x_2, z_2, E\Theta_1)\left(g_{2a}(x_2, z_2, \Theta_2) + g_{2b}(x_2, \Theta_2, Gw)\right)$$
$$+ u(G\hat{w}, \hat{z}_1, \Theta_1, x_2, \tilde{z}_2, E)\right) + h(Gw, z_1, \Theta_1, x_2, z_2, E)$$
$$\dot{E} = \left(Q(\tilde{z}_2) + E^{\mathsf{T}} - E\right)E \qquad (3.127)$$
$$\dot{e}_w = Se_w + l_w(E\Theta)$$
$$\dot{e}_z = -\Gamma e_z + Ge_w + l_z(E\Theta)$$
$$\dot{E}_\Theta = \left(Q\left(J_1^{-1}e_z\right) + L(E\Theta)\right)E\Theta + [Q\left(J_1^{-1}z_1\right), E\Theta],$$

where the observer gains l_w, l_z, L are chosen as in Lemma 3.20. In the following, we need the closed loop behavior in both the original coordinates and error coordinates. For convenience, we summarize the different state representations of the closed loop in Table 3.1.

Similar to the problem in Section 3.2, we cannot utilize the function $V_1 + V_2$ where V_1 is the Lyapunov function from the state feedback case and V_2 is the Lyapunov function for the observer to establish convergence. The reason is that the derivative $\dot{V}_1 + \dot{V}_2$ along solutions of the closed loop, i.e. (3.81), (3.82), the first equation in (3.79) and (3.127) with (3.126), contains indefinite expressions. Hence we cannot show convergence utilizing $V_1 + V_2$. Moreover, we cannot utilize cascade arguments due to the presence of multiple disconnected sets of equilibria.

To show convergence, we utilize once more the height functions just as in Section 3.2. In the present case the closed loop in error states plays the role of the vector field $\dot{x} = f(x)$ for

	components (original)	vector fields (original)	components (error)	vector fields (error)
$\mathcal{M}_1$	(w, z_1, Θ_1)	(3.82), (3.81)	(w, z_1, Θ_1)	(3.82), (3.81)
$\mathcal{M}_2$	(x_2, z_2, Θ_2)	(3.79), (3.126)	$(x_2, \tilde{z}_2, E)$	first eq. in (3.79) first two eq. in (3.127)
$\mathcal{M}_3$	$(\hat{w}, \hat{z}_1, \hat{\Theta}_1)$	(3.109)	$(e_w, e_z, \hat{\Theta}_1)$	last three eq. in (3.127)

Table 3.1: A summary of the original and error state representation of the closed loop. The closed loop state space is $\mathcal{M}_1 \times \mathcal{M}_2 \times \mathcal{M}_3$.

the height function. Therefore $\mathcal{M}$ is given by

$$\mathcal{M} = \mathcal{M}_1 \times \mathcal{M}_2 \times \mathcal{M}_3, \tag{3.128}$$

where $\mathcal{M}_1$ is defined by (3.83) with components (w, z_1, Θ_1), $\mathcal{M}_2$ by (3.80) with components $(x_2, \tilde{z}_2, E)$ and $\mathcal{M}_3$ by (3.110) with components $(((Ge_w)^\perp, Ge_w), e_z, E_\Theta)$. Lemma 3.20 shows that a part of the observer error typically converges, i.e. $(Ge_w(t; x_0), e_z(t; x_0), E_\Theta(t; x_0)) \stackrel{\text{typ}}{\to} (0, 0, I)$ for $t \to \infty$. Therefore, the component

$$\mathcal{N} = \mathcal{M}_1 \times \mathcal{M}_2 \times ((\mathbb{R}^{m-3}, 0), 0, I), \tag{3.129}$$

where the part $\mathbb{R}^{m-3}$ denotes the state space of $(Ge_w)^\perp$ is the only asymptotically stable invariant set for the observer error. In other words, $\mathcal{N}$ is the state space if the observer error is zero. The dynamics on $\mathcal{N}$ are given by (3.81), (3.82), $(G\hat{w})^\perp$, the first equation in (3.79) and the closed loop dynamics of the state feedback, i.e. (3.98) with (3.99). We know from the analysis in Lemma 3.17 that for almost all $x_0 \in \mathcal{N}$, we have $(\tilde{z}_2(t; x_0), E(t; x_0)) \to (0, I)$. This means that the solutions of the closed loop converge to $\mathcal{M}_1 \times (\mathbb{R}^l, 0, I) \times ((\mathbb{R}^{m-3}, 0), 0, I)$ as long as they are initialized on $\mathcal{N}$. However, even if we know that solutions typically converge towards $\mathcal{N}$ asymptotically, we do not know that generic closed loop solutions not initialized on $\mathcal{N}$ converge towards an equilibrium on $\mathcal{N}$, i.e. towards $\mathcal{E}$ given by

$$\mathcal{E} = \mathcal{M}_1 \times (\mathbb{R}^l, \mathcal{F}) \times ((\mathbb{R}^{m-3}, 0), 0, I), \tag{3.130}$$

where $\mathcal{F}$ is given by (3.100). In the following, we show that there is a height function which remedies the situation in our case. First we propose a height function.

Lemma 3.22. *Consider* $V_3 : \mathcal{M} \to \mathbb{R}$ *defined by*

$$V_3(w, z_1, \Theta_1, x_2, \tilde{z}_2, E, e_w, e_z, E_\Theta) = \frac{1}{2}\tilde{z}_2^\mathsf{T}\tilde{z}_2 + n - \mathrm{tr}(E), \tag{3.131}$$

where $\mathcal{M}$ is given by (3.128). V_3 is a height function for the pair $(\mathcal{N}, f)$ where $\mathcal{N}$ is given by (3.129) and f is the closed loop vector field in error coordinates.

Proof. The domain of V_3 is $\mathcal{M}$. As a consequence, $\mathcal{O} = \mathcal{M}$ where $\mathcal{O}$ is an open neighborhood of $\mathcal{N}$. If we evaluate V_3 on $\mathcal{N}$, we know that $Ge_w = 0$, $e_z = 0$ and $E = I$. Hence, $\dot{V}_3|_{\mathcal{N}}$ is the same as $\dot{V}_1$ in Lemma (3.17), thus we obtain $\dot{V}_3|_{\mathcal{N}} \leq 0$ and $\dot{V}_3|_{\mathcal{N} \setminus \mathcal{E}} < 0$. ∎

The following theorem contains the main result of this Section.

Theorem 3.23. *Assume that* $p : \mathcal{M}_2 \to \mathbb{R}^3$ *is linear in* z_2 *and that the* x_2-*dynamics are input-to-state stable with respect to* $Gw, \tilde{z}_2, \Theta_2$. *The certainty equivalence controller consisting of the observer* (3.109) *with the observer gains chosen as in Lemma 3.20 and the certainty equivalence implementation of the feedback* (3.126) *solves the Problem given in Problem statement 3.14.*

Proof. The proof is divided into three parts. In the first part we show that the assumptions to apply Theorem 1.17 are fulfilled. Therefore, we first show that the closed loop solutions remain bounded, which implies the property 3.14.c). Next we apply Theorem 1.17 to determine the sets that can possibly contain the ω-limit set of the solutions which converge to $\mathcal{N}$. In the final step we show that there is only one asymptotically stable component within these sets, thus proving the properties 3.14.a) from Problem statement 3.14. The invariance condition 3.14.b) from Problem statement 3.14 holds because of Corollary 3.18.

We first show that the solutions of the closed loop remain bounded for all initial conditions. The solution components are w, z_1, Θ_1 for the exosystem, $\hat{w}, \hat{z}_1, \hat{\Theta}_1$ for the observer and x_2, z_2, Θ_2 for the control system. The solution components w, z_1 remain bounded as explained after equation (3.82). The Θ_1-component is bounded since $\Theta_1 \in SO(3)$ and $SO(3)$ is compact, the same is true for the $\hat{\Theta}_1$ and the Θ_2-component. The observer error dynamics are determined by the last three equations in (3.127). With the chosen observer gains the convergence of the observer errors e_w, e_z, E_Θ is independent of the convergence of $\tilde{z}_2$ and of E. As a consequence we can utilize the proof for the convergence of the observer to guarantee boundedness of the respective states. More precisely, equation (3.125) guarantees that the errors e_w and e_z remain bounded. Since the w and the z_1-component of the closed loop solution remain bounded, this implies that the $\hat{w}$ and the $\hat{z}_1$ component remain bounded. The boundedness of the $\tilde{z}_2$-component (z_2-component) can be shown as follows. A calculation utilizing that p is linear in z_2 reveals

$$\dot{\tilde{z}}_2 = -\tilde{z}_2 + B(e_z, \tilde{z}_2, t) + a(e_z, Ge_w, t), \tag{3.132}$$

where $B(e_z, \tilde{z}_2, t) = Q(\tilde{z}_2)J_1^{-1}e_z$, $a(0,0,t) = 0$ and the dependence on t summarizes the remaining bounded functions. The solutions of this linear differential in $\tilde{z}_2$ exist for all times since the component e_z and Ge_w remain bounded. Furthermore, the component e_z and Ge_w typically converge to zero . Thus we see that $\tilde{z}_2$ remains bounded for all initial conditions. The x_2-component remains bounded in the closed loop since the x_2-dynamics are input-to-state stable with respect to Gw, z_2, Θ_2.

Now we apply Theorem 1.17. Lemma 3.22 shows that V_3 defined by (3.131) is a suitable height function for $(\mathcal{N}, f)$. As in Lemma 3.22, the set where $\dot{V}_3$ is zero is $\mathcal{E} = \mathcal{M}_1 \times (\mathbb{R}^l, \mathscr{F}) \times ((\mathbb{R}^{m-3}, 0), 0, I)$ and $\mathscr{F} = \{(\tilde{z}_2, E) \in \mathbb{R}^3 \times SO(3) | \tilde{z}_2 = 0, E^\top = E\}$. Lemma 3.24 shows that $\mathscr{F}$ is the set of critical points of V_3 and that V_3 is constant on the connected components of $\mathscr{F}$, thereby shows that the connected components $\mathcal{E}_k$ of $\mathcal{E}$ are contained in level sets of $V_3|_{\mathcal{N}}$. According to Theorem 1.17 the ω-limit of a closed loop solution that converges to $\mathcal{N}$ lies in a unique $\mathcal{E}_k$.

In the final step we analyze the stability of the connected components of $\mathcal{E} = \mathcal{M}_1 \times (\mathbb{R}^l, \mathscr{F}) \times ((\mathbb{R}^{m-3}, 0), 0, I)$ with $\mathscr{F} = \{(\tilde{z}_2, E) \in \mathbb{R}^3 \times SO(3) | \tilde{z}_2 = 0, E^\top = E\}$. For this step, we consider the closed loop system in error states (3.81), (3.82), the first equation in (3.79) and (3.126), (3.127) as a time-varying system with the states $\tilde{z}_2, E, Ge_w, e_z, E_\Theta$ and all other states from (3.126), (3.127), i.e. $(Gw)^\perp, z_1, \Theta_1, x_2$, as bounded time-varying functions. Then the stability analysis of the connected components of $\mathcal{E}$ in the closed loop system is equivalent to the stability analysis of the equilibria $\mathscr{F} \times (0,0,I)$. We abbreviate the state space of

93

$(\tilde{z}_2, E) \times (Ge_w, e_z, E_\Theta)$ by

$$\mathscr{L} = (\mathbb{R}^3, SO(3)) \times (\mathbb{R}^3, \mathbb{R}^3, SO(3)). \tag{3.133}$$

First we show asymptotic stability of the equilibrium $(0, I) \times (0, 0, I)$ for the time-varying system. This corresponds to the stability of the component $\mathscr{M}_1 \times (\mathbb{R}^l, 0, I) \times ((\mathbb{R}^{m-3}, 0), 0, I)$ in the closed loop. For the stability analysis, we utilize Theorem [62, Theorem 5] (see also Theorem 1.14). Consider the positive semidefinite function $V_4 : \mathscr{L} \to \mathbb{R}$ defined by $V_4 = V_2|_{\mathscr{L}}$, where V_2 is the function from Lemma 3.20. Lemma 3.20 implies (i) $V_4 \geq 0$ on $\mathscr{L}$, (ii) $V_4(\tilde{z}_2, E, Ge_w, e_z, E_\Theta) = 0$ if and only if $(Ge_w, e_z, E_\Theta) = (0, 0, I)$ and (iii) $\dot{V}_4 \leq 0$. Hence (i), (ii), (iii) imply that V_4 fulfills the first three conditions of Theorem [62, Theorem 5]. For the third condition we have to consider the dynamics of (3.127) on the set where V_4 is zero, i.e. on the set where $(Ge_w, e_z, E_\Theta) = (0, 0, I)$. These dynamics are given by (3.98) with (3.99), i.e. according to Lemma 3.17, the point $(0, I) \times (0, 0, I)$ is exponentially stable for (3.127) on the set where V_4 is zero. Hence Theorem [62, Theorem 5] implies that $(0, I) \times (0, 0, I)$ is uniformly stable. The attractivity of $(0, I) \times (0, 0, I)$ follows from the fact that this point is isolated in $\mathscr{L}$, hence we find a positively invariant neighborhood of the point (it is stable). According to Theorem 1.17 we converge to only one connected component of $\mathscr{F} \times (0, 0, I)$ in this neighborhood, and $(0, I) \times (0, 0, I)$ is the only component in the neighborhood, hence $(0, I) \times (0, 0, I)$ is attractive. Overall, $(0, I) \times (0, 0, I)$ is uniformly stable and attractive, hence asymptotically stable.

It is left to show that the remaining connected components $\mathscr{E}_k$ of $\mathscr{E}$ are unstable. These are given by $\mathscr{M}_1 \times (\mathbb{R}^l, E_0) \times ((\mathbb{R}^{m-3}, 0), 0, I)$ with $E_0 = E_0^\mathsf{T} \neq I$ (i.e. the equilibria $(0, E_0) \times (0, 0, I)$). Therefore we utilize again the dynamics of (3.127) on the set where V_4 is zero. Since this set is positively invariant, it is sufficient to show that these restricted dynamics are unstable. As mentioned before, these dynamics are given by (3.98) with (3.99), i.e. according to Lemma 3.17 all equilibria in $\mathscr{F}$ except $(0, I)$ are unstable. Hence the remaining connected components $\mathscr{E}_k$ are unstable and the proof is complete. ∎

Theorem 3.23 implies convergence of the closed loop system despite the presence of multiple connected components of equilibria. Note that the state feedback and the observer were essentially designed independently. The proposed state feedback and observer are only one possible choice. More precisely, under the mild assumption that the state feedback has an associated height function (as V_3 from Lemma 3.22 in our state feedback), and we have an typically convergent observer, then a controller consisting of the observer and a certainty equivalence implementation of the state feedback solves the problem from Problem statement 3.14. This establishes essentially a sort of separation principle.

3.3.6 Regulation problem for a satellite with movable appendages

In the following, we apply the theory from the Sections 3.3.3-3.3.5 to solve a tracking control problem for a satellite. We assume that the satellite consists of three rigid bodies, a solid cylindrical main body and two identical solar panels attached to the main body by massless rods, see Figure 3.7 for an illustration. We assume that the rods are perpendicular to the axis of rotational symmetry of the cylinder, that both rods are collinear and that the line through both rods passes through the center of mass of the cylinder. Furthermore, we assume that the movement of the solar panels relative to the main body is restricted to rotations around the rod and that the rod acts like a torsion spring damper system. The rods are designed in such a way that they damp possible movements of the panels, i.e. we assume that the damping is strong

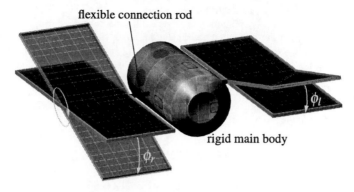

Figure 3.7: An illustration of the controlled satellite. The goal is that the main body asymptotically tracks a reference attitude irrespective of the panel movements and external disturbances. The picture is from Michalowsky [96] or Schmidt et al. [124] respectively.

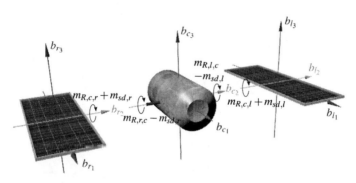

Figure 3.8: Free body diagram of the controlled satellite. We introduce four moments, the given moments $M_{sd,l}, M_{sd,r}$ due to the spring damper characteristics of the rod and the constraint moments $M_{R,l,c}, M_{R,r,c}, M_{R,c,l}, M_{R,c,r}$. The picture is from Michalowsky [96] or Schmidt et al. [124] respectively.

enough to keep the panel movements bounded. Since the solar panels are identical, the center of mass of the satellite coincides with the center of mass of the cylinder. We do not include assumptions on the actuators, e.g. momentum wheels, which would affect the momentum balance via the preservation of angular momentum Crouch [33], Sidi [135]. In the following, we derive a model of the form (3.79) for the satellite. Even though the modeling is straight forward, we present it here with all steps for the sake of completeness.

Modeling

First, we derive the kinematic equations of the system. Therefore, we introduce four orthonormal coordinate frames: an inertial frame $A = (a_1, a_2, a_3)$, a body fixed frame $B_c = (b_{c_1}, b_{c_2}, b_{c_3})$

for the main body and two body fixed frames $B_l = (b_{l_1}, b_{l_2}, b_{l_3})$, $B_r = (b_{r_1}, b_{r_2}, b_{r_3})$ for the left and the right panel, respectively. Without loss of generality, the body fixed frames are aligned with the principal axes of the associated bodies. Furthermore, we assume that b_{l_2} and b_{r_2} are collinear with the rods of the solar panels. See Figure 3.8 for an illustration. Let Θ_c denote the attitude of B_c relative to A. As mentioned in Section 3.3.1, we utilize the same convention as Murray et al. [102], i.e. Θ_c denotes the transformation between B_c and A in coordinates of A. The attitude of the left panel Θ_l^c and of the right panel Θ_r^c with respect to the frame B_c of the main body is given by a rotation around the axes b_{l_2} and b_{r_2}. Consequently, the attitudes of the panels with respect to the frame A are given by $\Theta_l = \Theta_c \Theta_l^c$ and $\Theta_r = \Theta_c \Theta_r^c$. The instantaneous spatial angular velocity of the main body with respect to the inertial frame A is given by

$$\omega_c(t) = Q^{-1}\left(\dot{\Theta}_c(t)\Theta_c^T(t)\right). \qquad (3.134)$$

Let $\gamma : \mathbb{R} \supset (-\varepsilon, \varepsilon) \to SO(3)$ be the attitude of the left panel, i.e. $\gamma(t) = \Theta_c(t)\Theta_l^c(t)$. Then the instantaneous spatial angular velocity ω_l of the left panel is

$$\begin{aligned} Q(\omega_l(t)) &= \dot{\gamma}(t)\gamma^T(t) \\ &= \dot{\Theta}_c(t)\Theta_c(t) + \Theta_c(t)\underbrace{\dot{\Theta}_l^c(t)\Theta_l^c(t)^T}_{=Q(\omega_l^c(t))}\Theta_c(t)^T, \end{aligned} \qquad (3.135)$$

where ω_l^c is the instantaneous angular velocity of the panel in the frame B_c. We utilize that for any $\Theta \in SO(3)$ and any $x \in \mathbb{R}^3$ we have the identity

$$Q(\Theta x)\Theta = \Theta Q(x), \qquad (3.136)$$

see e.g. [14, 3.10.1. xxxviii)]. Therefore, (3.136) implies for any $\Theta \in SO(3)$ the identity

$$Q(\omega)\Theta^T = Q\left(\Theta^T\Theta\omega\right)\Theta^T = \Theta^T Q(\Theta\omega). \qquad (3.137)$$

Consequently, we obtain for the left panel

$$Q(\omega_l(t)) = Q(\omega_c(t)) + Q(\Theta_c(t)\omega_l^c(t)). \qquad (3.138)$$

The angular velocities of the right panel are determined in the same way, i.e. we have

$$Q(\omega_r(t)) = Q(\omega_c(t)) + Q(\Theta_c(t)\omega_r^c(t)). \qquad (3.139)$$

The kinetic equations are determined by the rotational equations of motion of a rigid body. The rotational equations of motion of a rigid body are determined by Euler's second law. Relative to a point fixed in an inertial frame or relative to the center of mass we have

$$\widehat{J\omega} = M, \qquad (3.140)$$

where J is the inertia tensor in an inertial frame, ω is the instantaneous spatial angular velocity and M are the moments applied to the body expressed in an inertial frame. Here, the inertia tensors of the main body J_c, the left panel J_l and the right panel J_r in the inertial frame A are given by

$$\begin{aligned} J_c &= \Theta_c J_{c,0}\Theta_c^T \\ J_l &= \Theta_l J_{l,0}\Theta_l^T \\ J_r &= \Theta_r J_{r,0}\Theta_r^T. \end{aligned} \qquad (3.141)$$

The principal moments of inertia tensor $J_{c,0}$ of the main body is given by

$$J_{c,0} = \text{diag}\left(\left(\tfrac{1}{2}m_c r_c, \tfrac{1}{4}r_c^2 + \tfrac{1}{12}l_c^2, \tfrac{1}{4}r_c^2 + \tfrac{1}{12}l_c^2\right)\right), \tag{3.142}$$

which is the principal moments of inertia tensor of a cylinder with mass m_c, radius r_c and length l_c. The principal moments of inertia tensor of the left panel $J_{l,0}$ is given by

$$J_{l,0} = \tfrac{m_{lr}}{12} \text{diag}\left(\left(12l_{c,lr}^2 + l_{y,lr}^2 + l_{z,lr}^2, l_{x,lr}^2 + l_{z,lr}^2, 12l_{c,lr}^2 + l_{x,lr}^2 + l_{y,lr}^2\right)\right). \tag{3.143}$$

The principal moments of inertia tensor of the right panel is identical, i.e. $J_{r,0} = J_{l,0}$. Both panels are assumed to be cuboids of length $l_{x,lr}$, width $l_{y,lr}$ and height $l_{z,lr}$ and mass m_{lr}. The center of mass of $J_{l,0}, J_{r,0}$ has distance $l_{c,lr}$ to the center of mass of the main body. $J_{l,0}, J_{r,0}$ have the principal moments of inertia tensor of a cuboid plus the additional value given by Steiner's law, see e.g. Goldstein et al. [49].

In our case it is convenient to formulate Euler's second law in the inertial frame. Utilizing the free body diagram in Figure 3.8, we thus get

$$\begin{aligned}
\widehat{J_c \omega_c} + M_{sd,l} + M_{sd,r} - u_c - g_c(Gw) &= M_{R,l,c} + M_{R,r,c} \\
\widehat{J_l \omega_l} - M_{sd,l} &= M_{R,c,l} \\
\widehat{J_r \omega_r} - M_{sd,r} &= M_{R,c,r},
\end{aligned} \tag{3.144}$$

where $M_{sd,l}, M_{sd,r}$ are the given moments due to the spring damper characteristic of the rod, u_c is the system input, $M_{R,l,c}, M_{R,r,c}, M_{R,c,l}, M_{R,c,r}$ are constraint moments. In the following, we eliminate the constraint moments in (3.144). To eliminate the constraint moments we utilize the D'Alembert-Lagrange principle, which states for any virtual variation ξ and any constraint moment/force r that

$$\langle r, \xi \rangle = 0, \tag{3.145}$$

where $\langle x, y \rangle$ is the scalar product between x and y, see e.g. Lanczos [78], Arnold [7]. Virtual variations are movements of the mechanical system consistent with the constraint manifold. In case of a holonomic constraint this means that the virtual variations ξ are tangent vectors. In the case of the considered satellite this means

$$\langle Q(M_{R,l,c} + M_{R,r,c})\Theta_c, \Xi_c \rangle \;+\; \langle Q(M_{R,l,c})\Theta_l, \Xi_l \rangle \;+\; \langle Q(M_{R,l,c})\Theta_l, \Xi_l \rangle \;=\; 0, \tag{3.146}$$

where Ξ_c, Ξ_l, Ξ_r are tangent vectors compatible with the constraint manifold and the scalar product for $U_1\Theta, U_2\Theta \in T_\Theta SO(3)$ is given by $\langle U_1\Theta, U_2\Theta \rangle = \text{tr}\left((U_1\Theta)^\top (U_2\Theta)\right) = \text{tr}\left(U_1^\top U_2\right)$. The form of the virtual variations is determined by the kinematic equations (3.134), (3.138) and (3.139). More precisely, let $U_c = -U_c^\top, U_l = Q\left(\left(0\; u_l\; 0\right)^\top\right)$ and $U_r = Q\left(\left(0\; u_r\; 0\right)^\top\right)$ with $u_l, u_r \in \mathbb{R}$. Then

$$\begin{aligned}
\Xi_c &= U_c \Theta_c \\
\Xi_l &= (U_c + \Theta_c U_l)\Theta_l \\
\Xi_r &= (U_c + \Theta_c U_r)\Theta_r,
\end{aligned} \tag{3.147}$$

since the main body can rotate arbitrarily and the panels are restricted to rotations around their second axis which is collinear with the second main body axis. Therefore the D'Alembert-Lagrange principle results in

$$\begin{aligned}
&\left(\widehat{J_c \omega_c} + M_{sd,l} + M_{sd,r} - u_c - g_c(Gw)\right)^\top Q^{-1}(U_c) + \\
&\left(\widehat{J_l \omega_l} - M_{sd,l}\right)^\top Q^{-1}(U_c + \Theta_c U_l) + \\
&\left(\widehat{J_r \omega_r} - M_{sd,r}\right)^\top Q^{-1}(U_c + \Theta_c U_r) = 0.
\end{aligned} \tag{3.149}$$

Since the U_c, U_l, U_r are arbitrary and independent (variations), we obtain

$$\overrightarrow{J_c \omega_c} + \overrightarrow{J_l \omega_l} + \overrightarrow{J_r \omega_r} = u_c + g_c(Gw)$$
$$e_2^T \Theta_c^T \left(\overrightarrow{J_l \omega_l} - M_{sd,l} \right) = 0 \qquad (3.150)$$
$$e_2^T \Theta_c^T \left(\overrightarrow{J_r \omega_r} - M_{sd,r} \right) = 0$$

with $e_2 = \begin{pmatrix} 0 & 1 & 0 \end{pmatrix}^T$. The moments $M_{sd,l}$ and $M_{sd,r}$ due to the spring-damper characteristic of the rod, are given as functions of the relative angular velocities ω_l^c, ω_r^c (see (3.138) and (3.139)) and the angles of the left and right panel. Since the motion of the panels is one-dimensional, we define

$$\omega_l^c = \xi_l e_2, \qquad \omega_r^c = \xi_r e_2. \qquad (3.151)$$

The angles of the left and right panel are given by ϕ_l and ϕ_r. More precisely we have

$$\dot{\phi}_l = \xi_l$$
$$M_{sd,l} = -\Theta_c(d_l \xi_l + c_l \phi_l)e_2, \qquad (3.152)$$
$$\dot{\phi}_r = \xi_r$$
$$M_{sd,r} = -\Theta_c(d_r \xi_r + c_r \phi_r)e_2, \qquad (3.153)$$

where d_l, d_r, c_l, c_r are positive viscous damping and stiffness constants. Furthermore, define $\zeta_c = J_c \omega_c$. The states of the satellite are thus given by $\Theta_c, \zeta_c, \phi_l, \xi_l, \phi_r, \xi_r$. Therefore, the state space is a 10-dimensional manifold. In the following, we utilize the abbreviations given in (3.155). The overall control system is given by

$$\dot{\phi}_r = \xi_r$$
$$h^T A_r \dot{\zeta}_c + h^T b_r \dot{\xi}_r = -h^T \dot{A}_r \zeta_c - h^T \dot{b}_r - (d_r \xi_r + c_r \phi_r)$$
$$\dot{\phi}_l = \xi_l$$
$$h^T A_l \dot{\zeta}_c + h^T b_l \dot{\xi}_l = -h^T \dot{A}_l \zeta_c - h^T \dot{b}_l - (d_l \xi_l + c_l \phi_l) \qquad (3.154)$$
$$(I + A_l + A_r)\dot{\zeta}_c + b_l \dot{\xi}_l + b_r \dot{\xi}_r = -(\dot{A}_l + \dot{A}_r)\zeta_c$$
$$\qquad\qquad - \dot{b}_l \xi_l - \dot{b}_r \xi_r + u_c + g_c(Gw)$$
$$\dot{\Theta}_c = Q(\omega_c)\Theta_c,$$

where we utilized the following abbreviations

$$
\begin{array}{ll}
\dot{\Theta}_r^c = Q(\xi_r e_2)\Theta_r^c, & \dot{\Theta}_l^c = Q(\xi_r e_2)\Theta_l^c \\[4pt]
\dot{\Theta}_l = \dot{\Theta}_c \Theta_l^c + \Theta_c \dot{\Theta}_l^c, & \dot{\Theta}_r = \dot{\Theta}_c \Theta_r^c + \Theta_c \dot{\Theta}_r^c, \\[4pt]
J_l = \dot{\Theta}_l J_{l,0} \Theta_l^T + \Theta_l J_{l,0} \dot{\Theta}_l^T, & J_r = \dot{\Theta}_r J_{r,0} \Theta_r^T + \Theta_r J_{r,0} \dot{\Theta}_r^T, \\[4pt]
\overrightarrow{J_c^{-1}} = \dot{\Theta}_c J_{c,0}^{-1} \Theta_c^T + \Theta_c J_{c,0}^{-1} \dot{\Theta}_c^T, & \\[4pt]
h = \Theta_c e_2, & \dot{h} = Q(J_c^{-1} \zeta_c) h, & (3.155) \\[4pt]
A_l = J_l J_c^{-1}, & \dot{A}_l = \dot{J}_l J_c^{-1} + J_l \overrightarrow{J_c^{-1}}, \\[4pt]
b_l = J_l h, & \dot{b}_l = \dot{J}_l h + J_l \dot{h}, \\[4pt]
A_r = J_r J_c^{-1}, & \dot{A}_r = \dot{J}_r J_c^{-1} + J_r \overrightarrow{J_c^{-1}}, \\[4pt]
b_r = J_r h, & \dot{b}_r = \dot{J}_r h + J_r \dot{h}.
\end{array}
$$

Pertaining to the theoretical part of the Section, ζ_c would play the role of z_2, Θ_c the role of Θ_2 and the remaining states the role of x_2 in (3.79). The exosystem in the application scenario is the same as in the theoretical part, i.e.

$$
\begin{aligned}
\dot{w} &= Sw \\
\dot{z}_1 &= -\Gamma z_1 + Gw \\
\dot{\Theta}_1 &= Q\left(J_1^{-1} z_1\right)\Theta_1 \\
y_1 &= \Theta_1.
\end{aligned}
\tag{3.156}
$$

In the context of the application scenario, the exosystem is a reference system. Similar to the theoretical part, the objective is that Θ_c $(=\Theta_2)$ asymptotically tracks Θ_1.

Analysis

In the following, we verify briefly that all the necessary conditions to apply the theory developed in the previous sections are fulfilled. The property that T defined in (3.94) is a diffeomorphism can be verified directly. Since p from (3.79) is zero in the application scenario, $J_2^{-1} + \frac{\partial p}{\partial z_2}(x_2, z_2, \Theta_2) = J_2^{-1}$ $(= J_c^{-1})$ is non-singular. Since the x_2-part in the application scenario is given by the $(\phi_l, \xi_l, \phi_r, \xi_r)$-dynamics, we have to show that these are input-to-state stable with respect to the (Θ_c, ζ_c)-dynamics (inputs), see Corollary 3.18. For this, we transform the system from $(\zeta_c, \phi_l, \xi_l, \phi_r, \xi_r)$-states into $(\zeta_c, \phi_l, \eta_l, \phi_r, \eta_r)$-states by means of the transformation

$$
\begin{pmatrix}
\zeta_c \\ \phi_l \\ \xi_l \\ \phi_r \\ \xi_r
\end{pmatrix}
=
\begin{pmatrix}
I & 0 & 0 & 0 & 0 \\
0 & 1 & 0 & 0 & 0 \\
-\frac{h^T A_l}{h^T b_l} & 0 & 1 & 0 & 0 \\
0 & 0 & 0 & 1 & 0 \\
-\frac{h^T A_r}{h^T b_r} & 0 & 0 & 0 & 1
\end{pmatrix}
\begin{pmatrix}
\zeta_c \\ \phi_l \\ \eta_l \\ \phi_r \\ \eta_r
\end{pmatrix}.
\tag{3.157}
$$

The transformation aims to produce a block-triangular system form for the $(\zeta_c, \phi_l, \xi_l, \phi_r, \xi_r)$-part of (3.154). A lengthy calculation then shows that the dynamics of the $(\zeta_c, \phi_l, \eta_l, \phi_r, \eta_r)$-states are given by

$$
\begin{pmatrix}
I + A_l + A_r - b_l \frac{h^T A_l}{h^T b_l} - b_r \frac{h^T A_r}{h^T b_r} & 0 & b_l & 0 & b_r \\
\hline
0 & 1 & 0 & 0 & 0 \\
\hline
0 & 0 & j_l & 0 & 0 \\
\hline
0 & 0 & 0 & 1 & 0 \\
\hline
0 & 0 & 0 & 0 & j_r
\end{pmatrix}
\begin{pmatrix}
\dot{\zeta}_c \\ \dot{\phi}_l \\ \dot{\eta}_l \\ \dot{\phi}_r \\ \dot{\eta}_r
\end{pmatrix}
$$

$$
=
\begin{pmatrix}
I + A_l + A_r - b_l \frac{h^T A_l}{h^T b_l} - b_r \frac{h^T A_r}{h^T b_r} & 0 & b_l & 0 & b_r \\
\hline
-\frac{h^T A_l}{h^T b_l} & 0 & 1 & 0 & 0 \\
\hline
\dot{h}^T A_l + d_l \frac{h^T A_l}{h^T b_l} & -c_l & -d_l & 0 & 0 \\
\hline
-\frac{h^T A_r}{h^T b_r} & 0 & 0 & 0 & 1 \\
\hline
\dot{h}^T A_r + d_r \frac{h^T A_r}{h^T b_r} & 0 & 0 & -c_r & -d_r
\end{pmatrix}
\begin{pmatrix}
\zeta_c \\ \phi_l \\ \eta_l \\ \phi_r \\ \eta_r
\end{pmatrix}
+
\begin{pmatrix}
u^* \\ 0 \\ 0 \\ 0 \\ 0
\end{pmatrix}
\tag{3.158}
$$

$$
\dot{\Theta}_c = Q\left(J_c^{-1}\zeta_c\right)\Theta_c,
$$

with $u^* = u + g_c(Gw)$ and the abbreviations from (3.155). The block triangular form is apparent in the matrix on the left hand side of the $(\zeta_c, \phi_l, \eta_l, \phi_r, \eta_r)$-differential equation. To show the ISS property we consider the (ϕ_l, η_l)-dynamics, the reasoning for the (ϕ_r, η_r)-dynamics is analogous. From (3.158), we see

$$
\begin{pmatrix} 1 & 0 \\ 0 & (J_{l,0})_{22} \end{pmatrix} \begin{pmatrix} \dot{\phi}_l \\ \dot{\eta}_l \end{pmatrix} = \begin{pmatrix} 0 & 1 \\ -c_l & -d_l \end{pmatrix} \begin{pmatrix} \phi_l \\ \eta_l \end{pmatrix} + \begin{pmatrix} -\frac{h^{\mathsf{T}}A_l}{h^{\mathsf{T}}b_l} \\ h^{\mathsf{T}}A_l + d_l\frac{h^{\mathsf{T}}A_l}{h^{\mathsf{T}}b_l} \end{pmatrix} \zeta_c, \tag{3.159}
$$

where $(J_{l,0})_{22}$ is a positive constant. This shows that the (ϕ_l, η_l)-dynamics consist of an asymptotically stable linear system with inputs depending on the bounded variables $\Theta_c, \Theta_l^c, \Theta_r^c$ and on $\dot{h}, \zeta_c$. Since $\dot{h}^{\mathsf{T}} = -e_2^{\mathsf{T}}\Theta_c^{\mathsf{T}}Q(\omega_c)$, we see that $\dot{h}, \zeta_c$ depend only on the input $\Theta_c\omega_c = \zeta_c$, hence the (ϕ_l, η_l)-dynamics are input-to-state stable with respect to the (Θ_c, ζ_c)-dynamics. As a consequence, we can design the state feedback as explained in Section 3.3.3. Since the exosystem (3.156) is the same as the (3.81),(3.82), we can utilize the observer design from Section 3.3.4, where the observability condition for this system and the observer design are explained in detail. Overall, all conditions to design an output feedback controller with the proposed method are fulfilled. We present the controller design and some simulations in the following section.

Controller design

As explained in Section 3.3.5, the controller design is a two-step procedure. We start with the state feedback, which we design for the transformed control system (3.158). The design procedure for the state feedback is explained in Section 3.3.3 and is given by (3.97) and (3.99), i.e. the state feedback is determined by g_{2a}, g_{2b} and p in the considered problem. In the satellite problem, we have $p = 0$. To give the explicit equations for g_{2a} and g_{2b}, we introduce the abbreviations

$$
\begin{aligned}
C_c &= I + A_l + A_r - b_l\frac{h^{\mathsf{T}}A_l}{h^{\mathsf{T}}b_l} + b_r\frac{h^{\mathsf{T}}A_r}{h^{\mathsf{T}}b_r} \\
C_l &= \dot{h}^{\mathsf{T}}A_l + d_l\frac{h^{\mathsf{T}}A_l}{h^{\mathsf{T}}b_l} \\
C_r &= \dot{h}^{\mathsf{T}}A_r + d_r\frac{h^{\mathsf{T}}A_l}{h^{\mathsf{T}}b_l}.
\end{aligned} \tag{3.160}
$$

Then

$$
\begin{aligned}
g_{2a}(\phi_l, \eta_l, \Theta_l^c, \phi_r, \eta_r, \Theta_r^c, \zeta_c, \Theta_c) &= -\frac{b_l}{C_c(J_{l,0})_{22}}(C_l\zeta_c - c_l\phi_l - d_l\eta_l) + \frac{\dot{b}_l}{C_c}\eta_l \\
&\quad - \frac{b_r}{C_c(J_{r,0})_{22}}(C_r\zeta_c - c_r\phi_r - d_r\eta_r) + \frac{\dot{b}_r}{C_c}\eta_r, \\
g_{2b}(Gw, \Theta_l^c, \Theta_r^c, \Theta_c) &= \frac{g_c(Gw)}{C_c}.
\end{aligned} \tag{3.161}
$$

Furthermore, we have

$$h(Gw, z_1, \Theta_1, \zeta_c, E) = \widehat{J_c^{-1}\zeta_c} - Q^{-1}\left(EQ\,\widehat{(J_1^{-1}z_1)}\,E^{\mathsf{T}} - \dot{E}^{\mathsf{T}} - \dot{E}\right)$$

$$A(z_1, \Theta_1, E) = EQ\,(J_1^{-1}z_1)\,E^{\mathsf{T}} + E^{\mathsf{T}} - E \tag{3.162}$$

$$Q\left(\tilde{\zeta}_c\right) = Q\,(J_c^{-1}\zeta_c) - A(z_1, \Theta_1, E)$$

$$k(\tilde{\zeta}_c, E) = -\tilde{\zeta}_c.$$

The next step in the controller design is the observer design. In the application scenario we utilize the following parameters for the exosystem:

$$S = \begin{pmatrix} 0 & 2\pi & 0 & 0 \\ -2\pi & 0 & 0 & 0 \\ 0 & 0 & 0 & -20\pi \\ 0 & 0 & 20\pi & 0 \end{pmatrix}, \, G = \begin{pmatrix} 1 & 0 & 0 & 0 \\ 0 & 0 & 1 & 0 \\ 10 & 0 & 1 & 0 \end{pmatrix}, \tag{3.163}$$

$$\Gamma = 0, \, J_1 = \text{diag}([1, 2, 4]).$$

To obtain the observer gains given in Lemma 3.20 we have to solve the LMI feasibility problem given by (3.114) and (3.115). The matrices obtained as solutions to this problem with the given parameters are

$$P_0 = \begin{pmatrix} 1.17 & 0 & 0 & 0 \\ 0 & 2.21 & 0 & 0.01 \\ 0 & 0 & 1.14 & 0 \\ 0 & 0.01 & 0 & 1.14 \end{pmatrix},$$

$$\tag{3.164}$$

$$P_1 = \begin{pmatrix} 1.13 & 0 & -0.08 \\ 0 & 1.14 & 0 \\ -0.08 & 0 & 0.34 \end{pmatrix}, P_2 = \begin{pmatrix} 0 & 0.05 & 0 & 0.02 \\ 0 & 0 & 0 & 0.02 \\ 0 & 0.54 & 0 & 0 \end{pmatrix}.$$

With the given P_0, P_1, P_2, the observer is given by (3.109) where the observer gains l_w, l_z, L are given by (3.117), (3.118) and (3.119).

The controller is then given by the certainty equivalence implementation of the state feedback (3.126) using the estimated states of the observer (3.109). For the simulation of the closed loop we utilized the following parameters

$$m_c = 7000, \, l_c = 13.2, \, r_c = 2.1,$$

$$m_{lr} = 1000, \, l_{x,lr} = 7.56, \, l_{y,lr} = 2.46,$$

$$l_{z,lr} = 0.1, \, l_{c,lr} = 3.43,$$

$$c_l = 10000, \, c_r = 15000, \, d_l = 100, \, d_r = 400.$$

$$\tag{3.165}$$

The main body parameters utilized in the simulation are loosely motivated by the parameters of the Hubble space telescope[1]. The initial conditions for the simulations are

$$\xi_l(0) = \xi_r(0) = \phi_l(0) = \phi_r(0) = 0$$

$$z_2(0) = \begin{pmatrix} 0.94 \\ 0.92 \\ 0.41 \end{pmatrix}, \, \Theta_2(0) = \begin{pmatrix} 0.72 & -0.69 & -0.11 \\ -0.68 & 0.73 & -0.11 \\ -0.16 & 0 & 0.99 \end{pmatrix}. \tag{3.166}$$

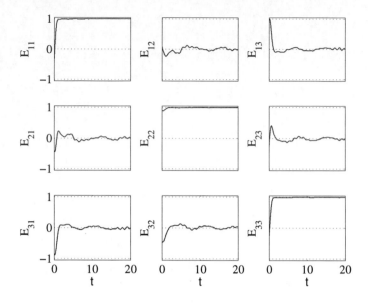

Figure 3.9: Exemplary time evolution of E.

The Figure 3.9 shows typical solutions for E, where the subplots correspond to the matrix entries. The convergence of the diagonal elements is usually much faster than the convergence of the off-diagonal elements.

3.3.7 Summary

In this Section we proposed a novel solution to a class of nonlinear output regulation problems which includes the rigid body attitude control problem. The presented controller is global and utilizes the natural attitude representation, i.e. $SO(3)$. We showed that it is possible to asymptotically stabilize the desired equilibrium in the zero error manifold and that the solutions typically converge to the desired equilibrium. The controller consists of a certainty equivalence implementation of a state feedback and an typically convergent observer. The design of the state feedback is essentially independent of the observer design, therefore the result establishes a separation principle for the rigid body equations. In contrast to many existing works, the presented design is relatively simple. More precisely, the utilized state feedback design builds on standard backstepping techniques, the observer design problem involves the solution of an LMI feasibility problem which can be computed efficiently using convex optimization algorithms. We demonstrated the controller design for a realistic application scenario of a satellite.

Future work could address the limitations of the present approach. More specifically, it would be desirable to cope with exosystems with unknown frequencies. Furthermore, it

---

[1] See e.g. the following website http://www.spacetelescope.org/about/general/fact_sheet/.

would be of interest to consider observer designs for the control systems which remove the assumption that the states of the control system are fully available from measurements.

3.4 Properties of some differential equations on $SO(n)$ and $SE(n)$

In this section we discuss the properties of a function and a differential equation on a smooth manifold. The results in Section 3.4 were made publicly available in Schmidt et al. [121]. More precisely, we consider the function $f : SO(n) \to \mathbb{R}$ and the differential equation $\dot{\Theta} = (\Theta^\mathsf{T} - \Theta)\Theta$ with $\Theta \in SO(n)$. This nonlinear differential equation appears throughout Section 3.2 and Section 3.3 as the error dynamics for the attitude error. In the following we show that this differential equation is a gradient flow and give a detailed analysis of its convergence properties. In the context of this section, we need the notions of *measure zero* and *dense* as given in Section 1.2.1.

Here, we consider a function and a differential equation on the set of special orthogonal matrices $SO(n) = \{\Theta \in \mathbb{R}^{n \times n} | \Theta^{-1} = \Theta^\mathsf{T}, \det(\Theta) = 1\}$. $SO(n)$ is a smooth manifold of dimension $\frac{n(n-1)}{2}$ with the subspace topology induced by $\mathbb{R}^{n \times n}$. The tangent space $T_\Theta SO(n)$ at Θ is given by

$$T_\Theta SO(n) = \{X \in \mathbb{R}^{n \times n} | X = \Omega\Theta, \ \Omega \in \mathbb{R}^{n \times n}, \ \Omega = -\Omega^\mathsf{T}\} \ , \tag{3.167}$$

see e.g. Helmke and Moore [58]. The Riemannian metric $g : T_\Theta SO(n) \times T_\Theta SO(n) \to \mathbb{R}$ induced by the standard Euclidean metric on $SO(n)$ is given by

$$g(\Omega_1\Theta, \Omega_2\Theta) = \mathrm{tr}\left((\Omega_1\Theta)^\mathsf{T}\Omega_2\Theta\right) = \mathrm{tr}\left(\Omega_1^\mathsf{T}\Omega_2\right), \tag{3.168}$$

see e.g. Helmke and Moore [58]. In the following, we define the differential and the Hessian of a function $f : SO(n) \to \mathbb{R}$ at a point $\Theta_0 \in SO(n)$. Let $\Gamma : (-\varepsilon, \varepsilon) \to SO(n)$ be a smooth curve with $\Gamma'(t) = \Omega(t)\Gamma(t)$, $\Gamma(0) = \Theta_0$ and $\dot{\Gamma}(0) = \Omega_0\Theta_0$ with $\Omega_0 \in \mathbb{R}^{n \times n}$ and $\Omega_0 = -\Omega_0^\mathsf{T}$. The differential $df_{\Theta_0} : T_{\Theta_0}SO(n) \to T_{f(\Theta_0)}\mathbb{R} \cong \mathbb{R}$ of a function $f : SO(n) \to \mathbb{R}$ at a point Θ_0 evaluated at $\Omega_0\Theta_0 \in T_{\Theta_0}SO(n)$ is defined by

$$df_{\Theta_0}(\Omega_0\Theta_0) = \tfrac{d}{dt}|_{t=0}(f \circ \Gamma)(t). \tag{3.169}$$

The *critical points of f* are the points Θ_0 where df_{Θ_0} is not surjective, see Guillemin and Pollack [51, p.22]. Because of $\dim(T_{f(\Theta_0)}\mathbb{R}) = 1$, this means that these are the points Θ_0 where $df_{\Theta_0} = 0$. The *gradient of f* is defined as the unique vector field $\mathrm{grad} f$ with

$$df_{\Theta_0}(\Omega_0\Theta_0) = g(\mathrm{grad} f(\Theta_0), \Omega_0\Theta_0), \tag{3.170}$$

see e.g. [81, Chapter 11]. The *Hessian $H_f(\Theta_0)$ of f* at a critical point Θ_0 evaluated at $(\Omega_0\Theta_0, \Omega_0\Theta_0)$ is defined by

$$H_f(\Theta_0)(\Omega_0\Theta_0, \Omega_0\Theta_0) = \tfrac{d^2}{dt^2}|_{t=0}(f \circ \Gamma)(t). \tag{3.171}$$

Since the Hessian at a critical point is bilinear and symmetric, we have for $(\Omega_1 + \Omega_2)\Theta_0 \in T_{\Theta_0}SO(n)$ with $\Omega_{1,2} = -\Omega_{1,2}^\mathsf{T}$ the equality

$$\begin{aligned} &H_f(\Theta_0)((\Omega_1 + \Omega_2)\Theta_0, (\Omega_1 + \Omega_2)\Theta_0) \\ &= H_f(\Theta_0)(\Omega_1\Theta_0, \Omega_1\Theta_0) + 2H_f(\Theta_0)(\Omega_1\Theta_0, \Omega_2\Theta_0) + H_f(\Theta_0)(\Omega_2\Theta_0, \Omega_2\Theta_0). \end{aligned} \tag{3.172}$$

As a consequence, the value $H_f(\Theta_0)(\Omega_1\Theta_0, \Omega_2\Theta_0)$ of H_f can be computed utilizing the values $H_f(\Theta_0)(\Omega_1\Theta_0, \Omega_1\Theta_0)$, $H_f(\Theta_0)(\Omega_2\Theta_0, \Omega_2\Theta_0)$, $H_f(\Theta_0)((\Omega_1 + \Omega_2)\Theta_0, (\Omega_1 + \Omega_2)\Theta_0)$ and (3.172). For this definition of the Hessian at a critical point and further details, see e.g. [58, Appendix C.5].

Lemma 3.24. *Consider the function* $f : SO(n) \to \mathbb{R}, \Theta \mapsto n - \mathrm{tr}(\Theta)$.

a) *The differential* df_{Θ_0} *of* f *at* Θ_0 *is given for any* $\Omega_0\Theta_0 \in T_{\Theta_0}SO(n)$ *by*

$$df_{\Theta_0}(\Omega_0\Theta_0) = -\frac{1}{2}\mathrm{tr}\left(\Theta_0(\Theta_0 - \Theta_0^\mathsf{T})\Omega_0\Theta_0\right)$$

and the critical points of f *are given by*

$$\mathcal{F} = \{\Theta_0 \in SO(n) | \Theta_0^\mathsf{T} = \Theta_0\}. \tag{3.173}$$

Furthermore, the gradient $\mathrm{grad}\, f(\Theta_0)$ *at* Θ_0 *is given by*

$$\mathrm{grad}\, f(\Theta_0) = \frac{1}{2}(\Theta_0 - \Theta_0^\mathsf{T})\Theta_0. \tag{3.174}$$

b) *The Hessian* $H_f(\Theta_0)$ *at a critical point* Θ_0 *is given by*

$$H_f(\Theta_0)(\Omega_1\Theta_0, \Omega_2\Theta_0) = \frac{1}{2}\mathrm{tr}\left(\Omega_1^\mathsf{T}\Theta_0\Omega_2 + \Omega_2^\mathsf{T}\Theta_0\Omega_1\right).$$

c) *The set of critical points* $\mathcal{F}$ *has the following properties:*

i) $\mathcal{F} = \cup_{k=0}^{\lfloor\frac{n}{2}\rfloor}\mathcal{F}_k$ *where*

$$\mathcal{F}_k = \{\Theta_0 \in SO(n) | \Theta_0 = \Theta_0^\mathsf{T}, \ \mathrm{tr}(\Theta_0) = n - 4k\}. \tag{3.175}$$

ii) *Each* $\mathcal{F}_k$ *is connected and isolated, i.e. there exists a neighborhood* U *of each* $\mathcal{F}_k$ *such that* $U \cap \mathcal{F}_l = \emptyset$ *for all* $l \neq k$.

iii) $\mathcal{F}_k$ *are compact submanifolds of dimension* $2k(n - 2k)$ *and the tangent space at* $\Theta_0 \in \mathcal{F}_k$ *is* $T_{\Theta_0}\mathcal{F}_k = \{\Sigma \in \mathbb{R}^{n\times n} | \Sigma = \Sigma^\mathsf{T}\} \cap T_{\Theta_0}SO(n)$.

iv) *For every* $k \in \{0,\ldots,\lfloor\frac{n}{2}\rfloor\}$ *and every* $\Theta_0 \in \mathcal{F}_k$ *we have*

$$\ker H_f(\Theta_0) = \{X \in T_{\Theta_0}SO(n)|$$
$$H_f(\Theta_0)(X,Y) = 0 \text{ for all } Y \in T_{\Theta_0}SO(n) \ \} = T_{\Theta_0}\mathcal{F}_k.$$

d) f *has a unique minimum at* $\Theta_0 = I$, *the other critical points are saddle points.*

Corollary 3.25. *The differential equation*

$$\dot{\Theta} = (\Theta^\mathsf{T} - \Theta)\Theta \tag{3.176}$$

is the gradient flow of $f : SO(n) \to \mathbb{R}, \Theta \mapsto 2n - 2\mathrm{tr}(\Theta)$ *with respect to the Riemannian metric* (3.168).

In the following we prove Lemma 3.24.

Proof. **a)** As stated above, $\Gamma : (-\varepsilon, \varepsilon) \to SO(n)$ is a differentiable curve with $\dot{\Gamma}(t) = \Omega(t)\Gamma(t)$, $\Gamma(0) = \Theta_0$ and $\dot{\Gamma}(0) = \Omega_0\Theta_0$ with $\Omega_0 \in \mathbb{R}^{n \times n}$ and $\Omega_0 = -\Omega_0^\mathsf{T}$. Then

$$
\begin{aligned}
df_{\Theta_0}(\Omega_0\Theta_0) &= \frac{d}{dt}\Big|_{t=0} (n - \mathrm{tr}\,(\Gamma(t))) = -\mathrm{tr}\,(\Omega_0\Theta_0) \\
&= -\frac{1}{2}\mathrm{tr}\,\left(\Theta_0^\mathsf{T}\Omega_0^\mathsf{T} + \Omega_0\Theta_0\right) \\
&= -\frac{1}{2}\mathrm{tr}\,\left((\Theta_0 - \Theta_0^\mathsf{T})\Omega_0\right) \\
&= -\frac{1}{2}\mathrm{tr}\,\left(\Theta_0^\mathsf{T}(\Theta_0 - \Theta_0^\mathsf{T})\Omega_0\Theta_0\right).
\end{aligned}
\tag{3.177}
$$

Therefore, the critical points of f are given by

$$
\{\Theta_0 \in SO(n) | \Theta_0 = \Theta_0^\mathsf{T}\}.
\tag{3.178}
$$

With the definition of the Riemannian metric by (3.168), the gradient $\mathrm{grad}\, f_\Theta$ at Θ_0 is given by

$$
\mathrm{grad}\, f(\Theta_0) = \frac{1}{2}(\Theta_0 - \Theta_0^\mathsf{T})\Theta_0^\mathsf{T}.
\tag{3.179}
$$

b) Let Θ_0 denote a critical point of f. As stated above, $\Gamma : (-\varepsilon, \varepsilon) \to SO(n)$ is a differentiable curve with $\dot{\Gamma}(t) = \Omega(t)\Gamma(t)$, $\Gamma(0) = \Theta_0$ and $\dot{\Gamma}(0) = \Omega_0\Theta_0$ with $\Omega_0 \in \mathbb{R}^{n \times n}$ and $\Omega_0 = -\Omega_0^\mathsf{T}$. Then

$$
\begin{aligned}
H_f(\Theta_0)(\Omega_0\Theta_0, \Omega_0\Theta_0) &= \frac{d^2}{dt^2}\Big|_{t=0}(f(\Gamma(t))) = -\mathrm{tr}\,\left(\ddot{\Gamma}(t)\right)\big|_{t=0} \\
&= -\mathrm{tr}\,\left(\dot{\Omega}(t)\Gamma(t) + \Omega(t)\dot{\Gamma}(t)\right)\big|_{t=0} \\
&= -\mathrm{tr}\,\left(\Omega_0^2\Theta_0\right) = \mathrm{tr}\,\left(\Omega_0^\mathsf{T}\Theta_0\Omega_0\right) = \mathrm{tr}\,\left(\Theta_0^\mathsf{T}\Omega_0^\mathsf{T}\Theta_0\Omega_0\Theta_0\right),
\end{aligned}
\tag{3.180}
$$

where we utilized that $\mathrm{tr}\,\left(\dot{\Omega}(0)\Theta_0\right) = 0$ since $\dot{\Omega}(t)$ is skew symmetric for all t and $\Theta_0 = \Theta_0^\mathsf{T}$ since Θ_0 is a critical point. Utilizing (3.172) we get

$$
\begin{aligned}
H_f(\Theta_0)(\Omega_1\Theta_0, \Omega_2\Theta_0) &= \frac{1}{2}\bigg(H_f(\Theta_0)((\Omega_1 + \Omega_2)\Theta_0, (\Omega_1 + \Omega_2)\Theta_0) \\
&\quad - H_f(\Theta_0)(\Omega_1\Theta_0, \Omega_1\Theta_0) - H_f(\Theta_0)(\Omega_2\Theta_0, \Omega_2\Theta_0)\bigg) \\
&= \frac{1}{2}\mathrm{tr}\,\left((\Omega_1 + \Omega_2)^\mathsf{T}\Theta_0(\Omega_1 + \Omega_2) - \Omega_1^\mathsf{T}\Theta_0\Omega_1 - \Omega_2^\mathsf{T}\Theta_0\Omega_2\right) \\
&= \frac{1}{2}\mathrm{tr}\,\left(\Omega_1^\mathsf{T}\Theta_0\Omega_2 + \Omega_2^\mathsf{T}\Theta_0\Omega_1\right).
\end{aligned}
\tag{3.181}
$$

c)i) Since Θ_0 is symmetric, Θ_0 is orthonormally diagonalizable, i.e. $\Theta_0 = \Pi^\mathsf{T} D\Pi$ for some diagonal D and orthonormal Π where the columns of Π are eigenvectors of Θ_0. Since $\Theta_0^\mathsf{T}\Theta_0 = I$, we get $D^2 = I$ and consequently the eigenvalues are ± 1. Since $\Theta_0 \neq I$ and $\det(\Theta_0) = 1$, we always have an even number of negative eigenvalues. A similarity transformation leaves the trace invariant, hence a critical point Θ_0 fulfills

$$
\mathrm{tr}\,(\Theta_0) = n - 4k,
\tag{3.182}
$$

where $k \in \{0, \ldots, \lfloor \frac{n}{2} \rfloor\}$ is the number of eigenvalue pairs which are -1.

c)ii) We start by showing that each $\mathscr{F}_k$ is path connected and thus connected. Let $k \in \{0, \ldots, \lfloor \frac{n}{2} \rfloor\}$ be arbitrary but fixed and let $\Theta_1, \Theta_2 \in \mathscr{F}_k$. Then there are orthogonal Π_1, Π_2 such that $\Theta_1 = \Pi_1^\mathsf{T} D \Pi_1$ and $\Theta_2 = \Pi_2^\mathsf{T} D \Pi_2$. Furthermore, there are real skew-symmetric matrices Ω_1, Ω_2 such that $\Pi_1 = \exp(\Omega_1)$ and $\Pi_2 = \exp(\Omega_2)$ with exp denoting the matrix exponential. Then $\alpha : [0,1] \to SO(n)$ defined by

$$t \mapsto \exp\left(\Omega_1^\mathsf{T}(1-t)\right) \exp\left(\Omega_2^\mathsf{T} t\right) D \exp\left(\Omega_2 t\right) \exp\left(\Omega_1(1-t)\right) \tag{3.183}$$

is a smooth curve in $\mathscr{F}_k$ which connects Θ_1 and Θ_2. Since $\Theta_1, \Theta_2 \in \mathscr{F}_k$ were arbitrary, this implies the path-connectedness of $\mathscr{F}_k$. To show that $\mathscr{F}_k$ is isolated, we utilize that $n - 4l = f|_{\mathscr{F}_l} \neq f|_{\mathscr{F}_k} = n - 4k$ for $l \neq k$. Then there is a $\varepsilon(l)$ with $(n - 4l - \varepsilon(l), n - 4l + \varepsilon(l)) \cap (n - 4k - \varepsilon(l), n - 4k + \varepsilon(l)) = \emptyset$ for $k \neq l$. As a consequence, the intersection of the preimage of these sets under f is empty. Since f is continuous and both, $(n - 4l - \varepsilon(l), n - 4l + \varepsilon(l))$ and $(n - 4k - \varepsilon(l), n - 4k + \varepsilon(l))$ are open, their preimages are open and contain $\mathscr{F}_l$ and $\mathscr{F}_k$ respectively. With $U_k(l) = f^{-1}\left((n - 4l - \varepsilon(l), n - 4l + \varepsilon(l))\right)$ we thus have $U_k(l) \cap \mathscr{F}_l = \emptyset$. Since this is possible for every $l \in \{0, \ldots, \lfloor \frac{n}{2} \rfloor\}$ and since a finite intersection of open sets is an open set, we find an open neighborhood U of $\mathscr{F}_k$ such that $U \cap \mathscr{F}_l = \emptyset$ for all $l \in \{0, \ldots, \lfloor \frac{n}{2} \rfloor\}$.

c)iii) The property that the $\mathscr{F}_k$ are submanifolds is given in [45]. The tangent space follows from (3.173).

c)iv) Let $k \in \{0, \ldots, \lfloor \frac{n}{2} \rfloor\}$ be arbitrary but fixed and $\Theta_0 \in \mathscr{F}_k$. Since $\ker H_f(\Theta_0) \supset T_{\Theta_0}\mathscr{F}_k$ is always true, we have to check $\ker H_f(\Theta_0) \subset T_{\Theta_0}\mathscr{F}_k$. Since every critical point is symmetric, there is an orthogonal Π such that $\Pi^\mathsf{T} \Theta_0 \Pi = D$ where D is a diagonal matrix with non-zero diagonal elements. Furthermore, we know that

$$
\begin{aligned}
H_f(\Theta_0)(\Omega_1\Theta_0, \Omega_2\Theta_0) &= -\frac{1}{2}\operatorname{tr}\left(\Omega_1^\mathsf{T}\Theta_0\Omega_2 + \Omega_2^\mathsf{T}\Theta_0\Omega_1\right) \\
&= -\frac{1}{2}\operatorname{tr}\left(\Theta_0^\mathsf{T}\Omega_1^\mathsf{T}\Theta_0\Omega_2\Theta_0 + \Theta_0^\mathsf{T}\Omega_2^\mathsf{T}\Theta_0\Omega_1\Theta_0\right) \\
&= -\frac{1}{2}\operatorname{tr}\left((\Omega_1\Pi^\mathsf{T} D\Pi)^\mathsf{T}\Pi^\mathsf{T} D\Pi(\Omega_2\Pi^\mathsf{T} D\Pi)\right) \\
&\quad -\frac{1}{2}\operatorname{tr}\left((\Omega_2\Pi^\mathsf{T} D\Pi)^\mathsf{T}\Pi^\mathsf{T} D\Pi(\Omega_1\Pi^\mathsf{T} D\Pi)\right) \\
&= -\frac{1}{2}\operatorname{tr}\left((\Pi\Omega_1\Pi^\mathsf{T} D)^\mathsf{T} D(\Pi\Omega_2\Pi^\mathsf{T} D)\right) \\
&\quad -\frac{1}{2}\operatorname{tr}\left((\Pi\Omega_2\Pi^\mathsf{T} D)^\mathsf{T} D(\Pi\Omega_1\Pi^\mathsf{T} D)\right) \\
&= -\frac{1}{2}\operatorname{tr}\left((\tilde{\Omega}_1 D)^\mathsf{T} D(\tilde{\Omega}_2 D) + (\tilde{\Omega}_2 D)^\mathsf{T} D(\tilde{\Omega}_1 D)\right) \\
&= H_f(D)(\tilde{\Omega}_1 D, \tilde{\Omega}_2 D)
\end{aligned}
$$

where $\tilde{\Omega}_1 = \Pi\Omega_1\Pi^\mathsf{T}$ and $\tilde{\Omega}_2 = \Pi\Omega_2\Pi^\mathsf{T}$. As a consequence

$$
\begin{aligned}
\ker H_f(\Theta_0) &= \{X \in T_{\Theta_0}SO(n) \,|\, H_f(\Theta_0)(X,Y) = 0 \text{ for all } Y \in T_{\Theta_0}SO(n)\} \\
&= \{X \in T_D SO(n) \,|\, \operatorname{tr}\left(X^\mathsf{T} D\tilde{\Omega}_2 D - D\tilde{\Omega}_2 DX\right) = 0 \text{ for all } \tilde{\Omega}_2 D \in T_D SO(n)\} \\
&= \{X \in T_D SO(n) \,|\, \operatorname{tr}\left(\tilde{\Omega}_2 D(X^\mathsf{T} - X)D\right) = 0 \text{ for all } \tilde{\Omega}_2 D \in T_D SO(n)\}.
\end{aligned}
$$

Observe that $\tilde{\Omega}_2$ is skew symmetric. Furthermore, $X^\mathsf{T} - X$ is skew symmetric and since D is diagonal with non-zero entries, $D(X^\mathsf{T} - X)D$ is skew symmetric. Because the equation

$\text{tr}\left(D\tilde{\Omega}_2 D(X^\mathsf{T} - X)\right) = 0$ has to hold for all skew-symmetric $\tilde{\Omega}_2$, we obtain $D(X^\mathsf{T} - X)D = 0$. With the non-singular D this implies $X^\mathsf{T} - X = 0$ which is equivalent to $X = X^\mathsf{T}$. In c)ii) we showed $T_{\Theta_0}\mathscr{F}_k$ is $\{\Sigma \in \mathbb{R}^{n\times n} | \Sigma = \Sigma^\mathsf{T}\} \cap T_{\Theta_0}SO(n)$, thus the previous calculation shows $\ker H_f(\Theta_0) \subset T_{\Theta_0}\mathscr{F}_k$.

d) Since $\Theta_0 = \Theta_0^\mathsf{T}$, Θ_0 is orthogonally diagonalizable, i.e. $\Theta_0 = \Pi^\mathsf{T}D\Pi$ for some diagonal D and orthogonal Π. Therefore

$$H_f((\Omega_0\Theta_0),(\Omega_0\Theta_0)) = \text{tr}\left(\Omega_0^\mathsf{T}\Theta_0\Omega_0\right) = -\text{tr}\left(\tilde{\Omega}_0^2 D\right), \tag{3.184}$$

where $\tilde{\Omega}_0 = \Pi\Omega_0\Pi^\mathsf{T}$ is skew symmetric. Consequently, $\text{tr}\left(\Omega_0^\mathsf{T}\Theta_0\Omega_0\right)$ is definite for all skew symmetric Ω_0 at a critical point Θ_0 if and only if $\text{tr}\left(\tilde{\Omega}_0^2 D\right)$ is definite for all skew symmetric $\tilde{\Omega}_0$ where $D = \Pi\Theta_0\Pi^\mathsf{T}$ is diagonal. Thus, we have to consider H_f only for diagonal D, i.e.

$$\begin{aligned}
H_f((\Omega_0\Theta_0),(\Omega_0\Theta_0)) &= -\text{tr}\left(\Omega_0^2 D\right) = -\sum_{1\leq k\leq n}\left(\Omega_0^2 D\right)_{kk} \\
&= -\sum_{1\leq k\leq n}(\Omega_0^2)_{kk}D_{kk} = -\sum_{1\leq k\leq n}\sum_{1\leq l\leq n}(\Omega_0)_{kl}(\Omega_0)_{lk}D_{kk} \\
&= \sum_{1\leq k,l\leq n}((\Omega_0)_{kl})^2 D_{kk} = \sum_{\substack{1\leq k,l\leq n \\ k\neq l}}((\Omega_0)_{kl})^2 D_{kk} \\
&= \sum_{1\leq k<l\leq n}(D_{kk}+D_{ll})((\Omega_0)_{kl})^2.
\end{aligned} \tag{3.185}$$

The critical points Θ_0 are such that $\Theta_0 \in SO(n)$ are symmetric, therefore all eigenvalues of Θ_0 are real. Observe now that $\Theta_0 = \Pi^\mathsf{T}D\Pi$ with orthogonal Π implies $D^2 = I$. Hence, the eigenvalues are $D_{kk} \in \{-1,1\}$ and since $\Theta_0 \in SO(n)$ the number of -1-eigenvalues is even. Consequently we have to determine the definiteness of H_f by considering (3.185) for all diagonal matrices D with ± 1 on the diagonals where the number of -1 entries is zero or even. Suppose first, that all D_{kk} are equal to 1, i.e. $D - I$. The associated Θ_0 is $\Theta_0 = \Pi^\mathsf{T}D\Pi = \Pi^\mathsf{T}\Pi = I$. (3.185) then implies $H_f(\Omega_0\Theta_0,\Omega_0\Theta_0) = \sum_{1\leq k<l\leq n}2(\Omega_0)_{kl}^2$, i.e. $H_f > 0$ for all skew-symmetric Ω_0. Thus H_f is positive definite if $\Theta_0 = I$. Suppose now there is an even number of eigenvalues D_{kk} equal to -1. Then, there are indices l,k and $l \neq k$ such that $D_{kk} = -1$ and $D_{ll} = -1$, and therefore there are skew symmetric Ω_0 such that $H_f(\Omega_0\Theta_0,\Omega_0\Theta_0) < 0$. As consequence, H_f is indefinite at a critical point Θ_0 where Θ_0 has an even number of negative eigenvalues $D_{kk} = -1$. Therefore, $\Theta_0 = I$ is the only local (global) minimum of f. All other critical points are saddle points. ∎

Definition 3.26. *[58, on p.21] Let $\mathscr{M}$ be a smooth Riemannian manifold and $f : \mathscr{M} \to \mathbb{R}$ be a smooth function. Denote the set of critical points of f by $C(f)$. f is called Morse-Bott function provided the following conditions are satisfied:*

a) f has compact sublevel sets.

b) $C(f) = \cup_{j=1}^k \mathscr{N}_j$ where $\mathscr{N}_j$ are disjoint, closed and connected submanifolds of $\mathscr{M}$ and f is constant on $\mathscr{N}_j$ for $j = 1,\ldots,k$.

c) $\ker H_f(x) = T_x\mathscr{N}_k$ for all $x \in \mathscr{N}_j$ and all $j = 1,\ldots,k$.

Lemma 3.27. *$f : SO(n) \to \mathbb{R}, \Theta \mapsto n - \text{tr}(\Theta)$ is a Morse-Bott function.*

Proof. We show only Definition 3.26a) since b) and c) were shown in Lemma 3.24. $SO(n)$ is compact, hence f attains its minimal and its maximal value on $SO(n)$. The minimal value of f is zero, the maximal value is $2n-2$ for n odd and $2n$ for n even. If n is odd we thus have

$$L_c = \{\Theta \in SO(n) | f(\Theta) \leq c\} = \begin{cases} f^{-1}([0,c]) & \text{for } c \leq 2n-2 \\ f^{-1}([0,2n-2]) & \text{for } c > 2n-2. \end{cases}$$

If n is even we have

$$L_c = \{\Theta \in SO(n) | f(\Theta) \leq c\} = \begin{cases} f^{-1}([0,c]) & \text{for } c \leq 2n \\ f^{-1}([0,2n]) & \text{for } c > 2n. \end{cases}$$

Since f is continuous, the preimage of a closed set is a closed set and since $SO(n)$ is bounded, its subsets are bounded as well. Since $SO(n) \subset \mathbb{R}^{n \times n}$, the boundedness and closedness of the sublevel sets implies their compactness. ∎

The convergence properties of the gradient flow associated with $\Theta \mapsto n - \text{tr}(\Theta)$ are thus determined by the following proposition.

Proposition 3.28. *[58, Proposition 3.9] Let $f : \mathcal{M} \to \mathbb{R}$ be a Morse-Bott function on a Riemannian manifold $\mathcal{M}$. The ω-limit set $\omega(x)$ of $x \in \mathcal{M}$ with respect to the gradient flow of f is a single critical point of f. Every solution of the gradient flow converges to an equilibrium point.*

To give a more detailed specification the convergence behavior of the gradient flow of $\Theta \mapsto n - \text{tr}(\Theta)$ we need the following result.

Lemma 3.29. *Let $\mathcal{M}$ be a smooth and compact Riemannian manifold of dimension m, $f : \mathcal{M} \to \mathbb{R}$ be a Morse-Bott function and denote the set of critical points of f by $C(f)$. Let $\mathcal{N}$ be a fixed connected component of $C(f)$ of dimension n. If at least one of the $m - n$ eigenvalues with nonzero real part of the linearization of $\text{grad} f$ at some $x \in \mathcal{N}$ has a real part greater than zero, then the set A of initial conditions $x_0 \in \mathcal{M}$ for which the solutions $t \mapsto \phi(t, x_0)$ of the gradient flow $\dot{x} = - \text{grad} f(x)$ converge towards $\mathcal{N}$, i.e.*

$$A = \{x_0 \in \mathcal{M} | \lim_{t \to \infty} \phi(t, x_0) \in \mathcal{N}\}, \tag{3.186}$$

has measure zero. Furthermore $\mathcal{M} \setminus A$ is dense in $\mathcal{M}$, i.e. $\overline{\mathcal{M} \setminus A} = \mathcal{M}$.

Proof. The goal of the proof is to show that A has measure zero and that $\mathcal{M} \setminus A$ is dense. We show this in the following way. First, we consider the set of points lying in a suitable neighborhood of $\mathcal{N}$ and which contains the orbits of the solutions of the gradient flow $\dot{x} = - \text{grad} f(x)$ which eventually converge towards $\mathcal{N}$. We utilize a result from [12] to conclude that this set has measure zero and $\mathcal{M}$ without this set is dense. Then, we utilize this set to derive the same result for A utilizing the properties of the flow of the gradient vector field on $\mathcal{M}$.

In the following, we apply [12, Proposition 4.1]. This proposition concerns the case of a a three times continuously differentiable vector field $v : \mathbb{R}^l \to \mathbb{R}^l$ together with submanifold of equilibria $\overline{\mathcal{N}}$ in $\mathbb{R}^l$ under the assumption that $\overline{\mathcal{N}}$ is normally hyperbolic with respect to v. Normal hyperbolicity of $\overline{\mathcal{N}}$ means that the linearization of the vector field v at $x \in \overline{\mathcal{N}}$ has $n - \dim \overline{\mathcal{N}}$ eigenvalues with real parts different from zero. Under these assumptions, there exists a neighborhood $\overline{\mathscr{U}}$ of $\mathscr{N}$ such that any solution $t \mapsto \phi(t,x_0)$ of $\dot{x} = v(x)$ with initial condition x_0 and with a forward orbit $\phi([0,\infty);x_0)$ in $\overline{\mathscr{U}}$ lies on the stable of manifold $W_{\text{loc}}^s(p)$ of a point $p \in \overline{\mathscr{N}}$. $W_{\text{loc}}^s(p)$ is defined by

$$W_{\text{loc}}^s(p) = \{x \in \mathscr{U} \,|\, \lim_{t \to \infty} \phi(t,x) = p\}. \tag{3.187}$$

We can always embed $\mathscr{M}$ into $\mathbb{R}^l$ for l large enough, see e.g. [81, Chapter 10], therefore we can utilize [12] also for our case of a vector field on a manifold $\mathscr{M}$.

Since $\mathscr{M}$ is compact and f is smooth, we have a global flow $\phi : \mathbb{R} \times \mathscr{M} \to \mathscr{M}$, which means that $t \mapsto \phi(t,x_0)$ is a solution of the gradient flow $\dot{x} = -\,\mathrm{grad}\,f(x)$ defined for all $t \in \mathbb{R}$ and with $\phi(0,x_0) = x_0$, see e.g. [81, Chapter 17]. Furthermore $\phi(t,\cdot) : \mathscr{M} \to \mathscr{M}$ is a diffeomorphism for every $t \in \mathbb{R}$. Since f is a Morse-Bott function, $\mathscr{N}$ is normally hyperbolic, i.e. the linearization of the gradient flow at any $x \in \mathscr{N}$ has exactly $m-n$ eigenvalues with real parts different from zero, see [94, p. 183, Morse-Bott functions]. According to [12, Proposition 4.1], we have a neighborhood $\mathscr{U}$ of $\mathscr{N}$ such that for every solution $\phi(\cdot,x)$ with $x \in \mathscr{N}$ and a forward orbit $\phi([0,\infty],x)$ in $\mathscr{U}$, the solution has to lie in one $W_{\text{loc}}^s(p)$ with $p \in \mathscr{N}$. We know from [13, Proposition 3.2], that if we choose $\mathscr{U}$ small enough, then the local stable manifold $W_{\text{loc}}^s(\mathscr{N})$ of $\mathscr{N}$ given by

$$W_{\text{loc}}^s(\mathscr{N}) = \cup_{p \in \mathscr{N}} W_{\text{loc}}^s(p) \tag{3.188}$$

is a smooth submanifold of dimension $m+k$ where $k < m-n$ is the number of eigenvalues with real part smaller than zero. Since the stable manifold $W_{\text{loc}}^s(\mathscr{N})$ is a submanifold of $\mathscr{M}$ with smaller dimension than $\mathscr{M}$, the stable manifold has measure zero and $\mathscr{M} \setminus W_{\text{loc}}^s(\mathscr{N})$ is dense in $\mathscr{M}$, see [81, Theorem 10.5].

Let A be defined by (3.186). Define A_1 by

$$A_1 = \{x \in A \,|\, \forall t \geq 1 : \phi(t,x) \in \mathscr{U}\} \tag{3.189}$$

and let A_k for $k \geq 2$ be defined by

$$A_k = \{x \in A \setminus (A_1 \cup \ldots \cup A_{k-1}) \,|\, \forall t \geq k : \phi(t,x) \in \mathscr{U}\}. \tag{3.190}$$

If $x \in A$, then there is an integer $k \in \mathbb{N}$ such that $x \in A_k$. As a consequence

$$A = \cup_{k \in \mathbb{N}} A_k. \tag{3.191}$$

Because of (3.190), $\phi(k,A_k) \subset \mathscr{U}$ for every A_k. Moreover, [12, Proposition 4.1] implies that

$$\phi(k,A_k) \subset W_{\text{loc}}^s(\mathscr{N}). \tag{3.192}$$

As subset of a set of measure zero, $\phi(k,A_k)$ has measure zero, see e.g. [81, Lemma A.60(b)]. Since $\phi(k,\cdot) : \mathscr{M} \to \mathscr{M}$ is a diffeomorphism, this means that A_k also has measure zero, see e.g. [81, Lemma 10.1]. According to (3.191), A is a countable union of the A_k, i.e. A is a countable union of sets of measure zero. Therefore, A has measure zero and as a consequence $\mathscr{M} \setminus A$ is dense.

∎

To finally derive the global stability properties of the identity matrix $I \in \mathbb{R}^{n \times n}$ for the gradient flow of $\Theta \mapsto n - \mathrm{tr}(\Theta)$ and thus also for the differential equation (3.176), we linearize the gradient flow around the equilibria.

Lemma 3.30. *The convergence properties of the gradient flow of $f : SO(n) \to \mathbb{R}, \Theta \mapsto n - \text{tr}(\Theta)$ are the following:*

a) *The ω-limit set of any solution is contained in the set of equilibria given by (3.173), i.e.*

$$\mathscr{F} = \{\Theta_0 \in SO(n)|\Theta_0^\top = \Theta_0\}.$$

b) *The equilibrium I is locally exponentially stable and all other equilibria are unstable.*

c) *The set of initial conditions for which the solutions of the gradient flow $\dot{x} = -\text{grad} f(x)$ of f converge towards I is dense in $SO(n)$ and the set of initial conditions for which the solutions of the gradient flow converge to the other equilibria has measure zero.*

Corollary 3.31. *The identity matrix I is an almost globally asymptotically stable equilibrium for the differential equation*

$$\dot{\Theta} = (\Theta^\top - \Theta)\Theta. \tag{3.193}$$

In the following we prove Lemma 3.30.

Proof. a) is a consequence of Lemma 3.27 and Proposition 3.28.

b) To prove the property b) we linearize the gradient flow with the vector field $\text{grad} f$ defined by $\Theta \mapsto \frac{1}{2}(\Theta^\top - \Theta)\Theta$ around the equilibria. To do this directly we compute $d\text{grad} f_{\Theta_0}(X) = \frac{d}{dt}|_{t=0}((\text{grad} f) \circ \Gamma)(t)$ where Θ_0 is an equilibrium and $\Gamma : (-\varepsilon, \varepsilon) \to SO(n)$ is smooth with $\Gamma(0) = \Theta_0$, $\dot{\Gamma}(0) = X$ and $X \in T_{\Theta_0}SO(n)$. This yields

$$\begin{aligned}
d\,\text{grad} f_{\Theta_0}(X) &= \frac{1}{2}\frac{d}{dt}|_{t=0}(\Gamma^\top(t) - \Gamma(t))\Gamma(t) \\
&= \frac{1}{2}\left(\dot{\Gamma}^\top(t)\Gamma(t) + \Gamma^\top(t)\dot{\Gamma}(t) - \dot{\Gamma}(t)\Gamma(t) - \Gamma(t)\dot{\Gamma}(t)\right)|_{t=0} \\
&= \frac{1}{2}(X^\top\Theta_0 + \Theta_0^\top X - X\Theta_0 - \Theta_0 X) \\
&= -\frac{1}{2}(\Theta_0 X + X\Theta_0),
\end{aligned} \tag{3.194}$$

since $\Theta_0 = \Theta_0^\top$ and $X = \Omega_0\Theta_0$ for a $\Omega_0 \in \mathbb{R}^{n \times n}$ with $\Omega_0 = -\Omega_0^\top$. Consequently the linearization of the gradient flow at an equilibrium Θ_0 is given by

$$\dot{X} = -\frac{1}{2}\left(\Theta_0^\top X + X\Theta_0\right), \tag{3.195}$$

where $X \in T_{\Theta_0}SO(n)$. Note that due to the simple nature of the Riemannian metric (3.168) and the connection of the linearization of a gradient flow to the Hessian, we could have obtained (3.195) directly from (3.180). More precisely, utilize

$$\begin{aligned}
H_f(\Theta_0)(X,X) &= \text{tr}\left(X^\top\Theta_0 X\right) = \text{vec}(\Theta_0 X^\top)\,\text{vec}(X) \\
&= \text{vec}(X)^\top(I \otimes \Theta_0 + (I \otimes \Theta_0)^\top)\,\text{vec}(X) \\
&= -\frac{1}{2}\text{vec}(X)^\top\,\text{vec}(\Theta_0^\top X + X\Theta_0).
\end{aligned} \tag{3.196}$$

If $\Theta_0 = I$, then the linearization is

$$\dot{X} = -X, \tag{3.197}$$

which shows that the equilibrium I is locally exponentially stable. Now consider the linearization at the other equilibrium points, i.e. $\Theta_0 \neq I$ and $\Theta_0 = \Theta_0^\mathsf{T}$. Since Θ_0 is symmetric, Θ_0 is orthonormally diagonalizable, i.e. $\Theta_0 = \Pi^\mathsf{T} D \Pi$ for some diagonal D and orthonormal Π where the columns of Π are eigenvectors of Θ_0. Since $\Theta_0^\mathsf{T} \Theta_0 = I$, we get $D^2 = I$ and consequently the eigenvalues are ± 1. Since $\Theta_0 \neq I$ and $\det(\Theta_0) = 1$, we always have an even number of negative eigenvalues with associated eigenvectors $v_1, \ldots, v_k$. Set $\overline{U} = v_1 v_2^\mathsf{T} - v_2 v_1^\mathsf{T}$ and $X = \overline{U}\Theta_0 \in T_{\Theta_0} SO(n)$. Therefore

$$
\begin{aligned}
-\frac{1}{2}(\Theta_0 X + X \Theta_0) &= -\frac{1}{2}\left(\Theta_0^\mathsf{T} \overline{U} \Theta_0 + \overline{U} \Theta_0 \Theta_0 \right) \\
&= -\frac{1}{2}\left(\Theta_0 (v_1 v_2^\mathsf{T} - v_2 v_1^\mathsf{T})\Theta_0 + (v_1 v_2^\mathsf{T} - v_2 v_1^\mathsf{T})\Theta_0 \Theta_0 \right) \\
&= -\frac{1}{2}\left((-v_1 v_2^\mathsf{T} + v_2 v_1^\mathsf{T})\Theta_0 + (-v_1 v_2^\mathsf{T} + v_2 v_1^\mathsf{T})\Theta_0 \right) = \overline{U}\Theta_0.
\end{aligned}
\tag{3.198}
$$

Therefore, $X = \overline{U}\Theta_0$ is an eigenvector of the operator defined by the right hand side of (3.195). Since the associated eigenvalue is positive (one), the linearization (3.195) is unstable. Consequently, the linearization of the gradient flow at the equilibria Θ_0 with $\Theta_0 \neq =I$ and $\Theta_0 = \Theta_0^\mathsf{T}$ is unstable, which proves b).

c) Denote the flow of $\dot{\Theta} = -\operatorname{grad} f(\Theta)$ by $\phi : \mathbb{R} \times SO(n) \to SO(n)$ and by B_k the set

$$
B_k = \{ x_0 \in SO(n) \,|\, \lim_{t \to \infty} \phi(t, x_0) \in \mathscr{F}_k \},
\tag{3.199}
$$

i.e. the set of initial conditions that converges to the connected component $\mathscr{F}_k$ of the set of critical points $\mathscr{F}$ given in Lemma 3.24c). Because of Proposition 3.28, we are certain that any solution of the gradient flow converges to the critical set of f, and as a consequence, $SO(n) = \cup_{k=0}^{\lfloor \frac{n}{2} \rfloor} B_k$. Then B_0 is the set of initial conditions for which the flow converges to I and $B = \cup_{k=1}^{\lfloor \frac{n}{2} \rfloor} B_k$ is the set of initial conditions for which the flow converges to any of the other critical points. In Lemma 3.29 we showed that B_k has measure zero and that $SO(n) \setminus B_k$ is dense in $SO(n)$. Since B is the union of a finite number of sets of measure zero, it has measure zeros, see e.g. [81, Lemma 10.1]. In particular $B_0 = SO(n) \setminus B$ is dense. ∎

3.5 Discussion

In this section, we considered two output regulation problems where the geometry of the state space poses a challenge to established design procedures. More specifically, the presence of a single asymptotically stable equilibrium implies the existence of additional equilibria. As a consequence, it is not possible to achieve global stability for the desired equilibrium point. We proposed an observer-based two step design procedure to solve the considered class of output regulation problems. The two step design procedure can be briefly summarized as follows: In the first step, find a state feedback which renders the desired equilibrium almost globally asymptotically stable in the closed loop consisting of the control system and the state feedback. Associated with the error system in the state feedback case is a height function. In the second step, we design an observer which converges typically. The output feedback is then given by a certainty equivalence implementation of the state feedback. To guarantee the convergence of the closed loop one has to ensure two properties. The first property is the boundedness of the closed loop solutions, which can be inferred with Lyapunov techniques. The second property is that the limit sets and their stability in the closed loop with the output feedback are the same as the limit sets and their stability in the state feedback case. We show the second property by utilizing the height function associated with the state feedback. Since the height function is independent of the feedback and the observer, we can change the observer and the state feedback as long as we can guarantee boundedness and as long as the technical properties of the height function with respect to the closed loop are fulfilled and still obtain the desired convergence properties. Furthermore, the design of the observer and the state feedback were independent. As a consequence, we obtain a separation principle for the considered problem class. Even though our presentation was given for the specific system classes on $SE(n)$ and for the rigid body problems, the principle approach and utilization of the height function are not restricted to this problem class.

In addition to the output regulation problems we considered the convergence properties of a differential equation on $SO(n)$. This differential equation appears in the error dynamics in the state feedback case for the considered output regulation problems. We show that the differential equation is a gradient flow of a Morse-Bott function, which implies the convergence of all solutions to a point in the set of equilibria. We show that there is only one asymptotically stable equilibrium and the other equilibria are unstable. In addition to that, we show that the set of initial conditions that converges to the unstable equilibria has measure zero, thereby establishing convergence for almost all initial conditions to the stable equilibrium.

4

Conclusions

The goal of this thesis was to obtain non-local or global analysis and synthesis results for synchronization problems and global output regulation problems for systems on $SE(n)$. In the following we summarize the obtained results in the light of this goal and give an outlook for possible future research directions.

4.1 Summary and outlook for the global output regulation problems

One of the main contributions of this thesis is a new separation principle for the class of control systems on $SE(n)$ and for the class of rigid body control systems as presented in Chapter 3. The established separation principle is a consequence of the proposed two-step design method. This method consists of a state feedback and an independent observer design and results in a typically convergent closed loop for a certainty equivalence implementation of the feedback. In other words, the closed loop has the property, that every solution converges to one of the isolated components of the set of equilibria, that the desired equilibrium is the only asymptotically stable equilibrium and that the other equilibria are unstable. From the proven statements we conclude that as long as the closed loop stays bounded and the properties of the height function associated to a stabilizing state feedback are fulfilled, we can utilize any certainty equivalence implementation of a typically stabilizing state feedback together with an observer and obtain typical convergence in the above sense for the certainty equivalence feedback.

As a consequence, our approach is analogous to the separation result from linear systems theory. There, the stability properties of the closed loop equilibrium in the case of the certainty equivalence implementation of the feedback is guaranteed for independently assigned poles for the observer and the feedback. Furthermore, our proofs show that the separation principle is not restricted to the considered case. Generally speaking, in a control system where the system together with a stabilizing state feedback has multiple equilibria and where we find a height function that fulfills the properties of the height function theorem (Arsie and Ebenbauer [9, Theorem 6]), a typically convergent observer will be sufficient for similar convergence properties of a certainty equivalence implementation of the state feedback, provided that the closed loop solutions remain bounded. Therefore our result can be considered as a rather general system theoretic result for nonlinear output feedback design in the presence of multiple equilibria.

One important factor in the applicability of output regulation problems for practical purposes is the exosystem. If the exosystem can produce a larger class of signals which can be handled with the respective output regulation approach, then the class of disturbances that can be suppressed and the class of references that can be followed is also larger. In the approach we presented, we assumed an exosystem which produces harmonic signals where we know all frequencies in the Fourier decomposition. For application purposes like the suppression of harmonic vibrations, it would be useful to remove the assumption that the frequencies of the Fourier decomposition of the exosystem signals are known. In this case the exosystem would be nonlinear, which is an additional challenge for the control design. A possible approach towards a solution to this problem should first test whether the existing internal model based approaches for the semiglobal problem like Serrani et al. [132] or the adaptive approaches like Marino and Tomei [89] are helpful for such an extension. Another interesting class of exosystems is one that has the ability to generate splines, see e.g. Cox et al. [32]. Exosystems which generate splines would be especially of interest for tracking applications, e.g. trajectory planning tasks in robotics where splines are used, see e.g. Biagiotti and Melchiorri [17]. Furthermore it would be of interest to consider exosystems that model the nonlinear vibrations present in flexible structures. This could be of interest for vibration suppression, which is an important topic in the control of flexible mechanical systems. Considerations in these direction would require an additional research effort concerning the controller design approach chosen in this thesis.

As mentioned in the introduction of Chapter 3, the two-step design approach we chose to solve the output regulation problem is not an internal model based approach in the sense of Isidori. The internal model based approaches developed for example in Isidori et al. [68] offer a systematic approach to achieve robust output regulation. Robustness in this context means robustness towards parametric uncertainties. From a theoretical point of view it would be interesting to investigate whether the techniques utilized in this thesis are also helpful for the internal model based design approaches in the sense of Isidori. A possible benefit would be the extension of the semi-global internal model based approaches to global approaches. This would remove the requirement to guarantee the invariance of operation regimes, see e.g. Isidori et al. [68], which is theoretically difficult for nonlinear systems and which complicates the implementation of such controllers. Approaches that yield controllers which are easier to implement are especially of interest from an application perspective.

4.2 Summary and outlook for synchronization problems

4.2.1 Synchronization for Lyapunov oscillators

An important theoretical basis for our separation results in the $SE(n)$ case were the height functions. These allowed us to determine the ω-limit sets of the solutions. The application of the height functions proved to be equally helpful for the analysis of synchronization in networks of Lyapunov oscillators in Chapter 2. We characterized synchronization in oscillator networks by two conditions, the ω-limit set property for synchronization (Lemma 2.42 a)) and the asymptotic phase property (Lemma 2.42 b)) for synchronization. Our analysis showed that a coupling in a network of Lyapunov oscillators that achieves attractivity and invariance of the synchronization manifold also guarantees the ω-limit set property, i.e. the solutions in the oscillator network converge to the synchronous orbit. Whether the existence of a Lyapunov function for the periodic orbits is restrictive or not depends on the existence of converse

Lyapunov theorems for this situation. For exponentially stable orbits such theorems exists, see e.g. Hauser [57]. For the other cases the converse Lyapunov theorems for closed sets might prove helpful, see e.g. Wilson [148]. The application of the height functions does not require assumptions on the convergence speed towards the synchronization manifold. As a consequence, in the cases where a Lyapunov function for the periodic orbits exists, a design guaranteeing invariance and attractivity of the synchronization manifold should also guarantee the convergence to the synchronous orbit, i.e. the ω-limit set property for synchronization.

In contrast to the ω-limit set property for synchronization, which we could guarantee by requiring the Lyapunov property for the single oscillators, we guaranteed the asymptotic phase property for synchronization in Chapter 2 by requiring exponential orbital stability of the synchronous orbit in the oscillator network. Requiring exponential orbital stability means that the characteristic exponents of the linearization about the synchronous solution have negative real part. The relation between the coupling in the oscillator network and the properties of the real part of the characteristic exponents is nonlinear, non-convex and depends on the exact knowledge of the synchronous solution. Therefore it is difficult to find a feedback design method which guarantees exponential orbital stability of the synchronous solution.

As a consequence of our findings, future research for synchronization in oscillator networks could concentrate on several problems. To better assess the principal possibilities of our approach to guarantee the ω-limit set property for synchronization, it would be useful to consider converse Lyapunov theorems for invariant sets in literature and study their applicability for periodic orbits. Equally important for our approach are investigations concerning conditions that guarantee the asymptotic phase property for synchronization. As mentioned before, one approach would be the investigation of the connection between the characteristic exponents associated with the synchronous solution and the coupling functions. This problem would also be of interest in a more general context, i.e. the design of feedback controllers to stabilize periodic orbits for nonlinear control systems. Results on this problem for high dimensional systems are rare in literature. Furthermore, research for the asymptotic phase property in dimensions higher than two is not complete, see especially Dumortier [37]. In addition to these two directions, another possibility would be the investigation of other phenomena which are observed in oscillator networks utilizing the same methods as in this work. More specifically, it is well known that many more effects than the frequency synchronization can be observed in oscillator networks. We mentioned that instead of frequency synchronization, Hoppensteadt and Izhikevich [59] utilizes the term entrainment, which is motivated by the observation that for a synchronous solution, the asymptotic frequencies of the oscillators in a network of oscillators have a fixed integer ratio. In the frequency synchronization case, this relation is 1 : 1 for every pair of oscillator in the network, but in principal other ratios are possible. The stability of these other frequency locked phenomena could be investigated in a similar fashion as the frequency synchronization phenomena.

4.2.2 Synchronization in Kuramoto models

In addition to our considerations of synchronization in state space models, we also analyzed the impact of delays on the synchronization effects in Kuramoto models. The conditions that we give and that guarantee synchronization for this model class always involve a region of attraction for the synchronous solution. As mentioned in Remark 2.25, our results for the delayed case imply similar results for the undelayed case, including the property that a region of attraction is involved. These results show that the classical stability analysis techniques utilized in control theory help to improve the insight for the synchronization effects in phase models.

Simulations give the impression that the region of attraction of synchronous solutions is rather large, i.e. if the coupling is strong enough the synchronous solution is the only stable solution for the exception of a "thin" set. As a consequence, in order to understand synchronization in phase models better more elaborate global analysis results are necessary. To give a more specific example about the possible benefits, we want to mention that it is well known that some phase models are gradient flows of differentiable functions, see e.g. Hoppensteadt and Izhikevich [59, Theorem 9.15]. In that case, the solutions converge to the set of critical points of the function whose gradient determines the vector field. For phase models, this function often has the property that it only depends on the phase difference between coupled oscillators. As a consequence, the critical points of these function form subsets of the state space. If the function is a Morse-Bott function, then results like Lemma 3.29 could establish almost global convergence to the (phase) synchronized state. Similar results for frequency synchronized solutions would be of interest. However, the analysis of frequency synchronization and more general frequency locking phenomena is more difficult, since the associated vector fields for established models like the Kuramoto model are no gradient vector fields. Just as in the case of the rotation matrices, an embedding of the phase model into a higher dimensional space might offer interesting analysis possibilities.

A

Rigid body motion

Here we shortly reconsider the dynamics of rigid bodies in the notation utilized in the thesis. Thorough and detailed expositions of the subject are given for example in Landau and Lifshitz [79], Arnold [7], Marsden and Ratiu [90], Goldstein et al. [49], Scheck [118] or, from a control point of view, Murray et al. [102]. Our exposition utilizes mainly arguments from Arnold [7] and Murray et al. [102].

A rigid body is a system of point masses or a system with a continuous mass distribution in three-dimensional Euclidean space with the holonomic constraint that the distance between any two points is constant. Hence, the configuration manifold of a rigid body is determined by all orientation preserving maps which leave the distance between two points invariant, i.e. the orientation preserving isometries of the real three-dimensional Euclidean space. The orientation preserving isometries of a finite-dimensional Euclidean space are characterized by the composition of a rotation and a translation, see e.g. Artin [10, Chapter 4, (5.20) Proposition] or the original result by Mazur and Ulam [93]. Consequently, we can represent these isometries in the rigid body case as an element of $\mathbb{R}^3 \times SO(3)$, see e.g. Arnold [7, Section 28A, Theorem].

The orientation preserving isometries are a transformation group and the operation defined by $((b_1, \Theta_1), (b_2, \Theta_2)) \mapsto (b_1 + \Theta_1 b_2, \Theta_1 \Theta_2)$ defines the associated product of the group elements in the representation $\mathbb{R}^3 \times SO(3)$. The set $SE(3) = \mathbb{R}^3 \times SO(3)$ with this product is called the *three dimensional special Euclidean Group*. If the base set is the n-dimensional Euclidean space for an arbitrary $n \in \mathbb{N}$, then the same operation is a product for the orientation preserving isometries in the representation $\mathbb{R}^n \times SO(n)$. The set $SE(n) = \mathbb{R}^n \times SO(n)$ with this product is called the *n-dimensional special Euclidean group*. Since $SO(n)$ is a smooth manifold, see e.g. Guillemin and Pollack [51, p.22], and the product of smooth manifolds is a smooth manifold, see e.g. Guillemin and Pollack [51, Chapter 1, §1], the set $SE(n) = \mathbb{R}^n \times SO(n)$ is a smooth manifold for $n \in \mathbb{N}$.

In the following, we discuss the kinematics of rigid body motion, i.e. the characteristics of the motion. As discussed in the previous paragraph, the configuration space of the rigid body motion is the manifold $SE(3)$. In other words, motions of the rigid body determine curves in $SE(3)$. To find the differential equations which describe the motion, we consider the coordinate description of a point on a rigid body. More precisely, let $A = (a_1, a_2, a_3)$ be a given right-handed orthonormal inertial coordinate frame and $C = (c_1, c_2, c_3)$ a right-handed orthonormal coordinate frame which is fixed relative to the rigid body, see Figure A.1. We denote the coordinate functions of the A and the C frame by κ_A and κ_C respectively, and the vector from the origin of A to the origin of C by p. Consider a point q fixed with respect to the

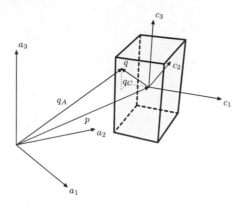

Figure A.1: An illustration of the coordinate frames which are utilized to describe the motion of a rigid body. We have two right-handed orthonormal frames. An inertial frame A given by the basis (a_1, a_2, a_3) and a body fixed frame C given by the basis (c_1, c_2, c_3). q denotes a fixed point of the rigid body. q_A is the vector to the point in the inertial frame, q_C the vector to the point in the body fixed frame.

rigid body and which is part of the rigid body. Denote by q_A the vector from the origin of A to q and by q_C the vector from the origin of C to q. Then, $p + q_C = q_A$ or, equivalently, $q_C = q_A - p$. The transformation matrix from C to A coordinates is $\Theta = \big(\kappa_A(c_1), \kappa_A(c_2), \kappa_A(c_3)\big)$. Since A and C are orthonormal frames, $\Theta \in SO(3)$. We utilize the same convention as Murray et al. [102] for the rotation matrix Θ, i.e. Θ denotes the transformation between the fixed body frame and an inertial frame in coordinates of the inertial frame. Then, we have

$$\Theta \kappa_C(q_C) = \kappa_A(q_A - p) = \kappa_A(q_A) - \kappa_A(p), \tag{A.1}$$

since the coordinate functions are linear. Then $g = (\kappa_A(p), \Theta) \in SE(3)$ determines the relationship between the two coordinate descriptions via

$$\kappa_A(q_A) = \Theta \kappa_C(q_C) + \kappa_A(p). \tag{A.2}$$

It is convenient to map $\kappa_A(q_A)$ to $(\kappa_A(q_A), 1)^\mathsf{T}$. Then the relation (A.2) can be written as a matrix multiplication, i.e.

$$\begin{pmatrix} \kappa_A(q_A) \\ 1 \end{pmatrix} = \begin{pmatrix} \Theta & \kappa_A(p) \\ 0 & 1 \end{pmatrix} \begin{pmatrix} \kappa_C(q_C) \\ 1 \end{pmatrix}. \tag{A.3}$$

$\kappa_A(q_A)$ is the position of the point q in spatial coordinates. If q moves as the rigid body moves, i.e. $q : \mathbb{R} \to \mathbb{R}^3$ is a differentiable map, then the relation (A.3) holds at each timepoint of the motion, i.e.

$$\begin{pmatrix} \kappa_A(q_A)(t) \\ 1 \end{pmatrix} = \begin{pmatrix} \Theta(t) & \kappa_A(p)(t) \\ 0 & 1 \end{pmatrix} \begin{pmatrix} \kappa_C(q_C) \\ 1 \end{pmatrix}. \tag{A.4}$$

The coordinates of q with respect to the frame C remain fixed, i.e. $\kappa_C(q_C)$ does not depend on time. As a consequence, we obtain an expression for the velocity $v_A(t) = \frac{d}{dt}\kappa_A(q_A)(t)$ of the point q in spatial coordinates by differentiation, i.e.

$$\begin{pmatrix} v_A(t) \\ 0 \end{pmatrix} = \frac{d}{dt}\begin{pmatrix} \kappa_A(q_A)(t) \\ 1 \end{pmatrix} = \begin{pmatrix} \dot{\Theta}(t) & \overrightarrow{\kappa_A(p)}(t) \\ 0 & 0 \end{pmatrix}\begin{pmatrix} \kappa_C(q_C) \\ 1 \end{pmatrix}$$

$$= \begin{pmatrix} \dot{\Theta}(t) & \overrightarrow{\kappa_A(p)}(t) \\ 0 & 0 \end{pmatrix}\begin{pmatrix} \Theta^{\mathsf{T}}(t) & -\Theta^{\mathsf{T}}(t)\kappa_A(p)(t) \\ 0 & 1 \end{pmatrix}\begin{pmatrix} \Theta(t) & \kappa_A(p)(t) \\ 0 & 1 \end{pmatrix}\begin{pmatrix} \kappa_C(q_C) \\ 1 \end{pmatrix}$$

$$= \begin{pmatrix} \dot{\Theta}(t)\Theta^{\mathsf{T}}(t) & -\dot{\Theta}(t)\Theta^{\mathsf{T}}(t)\kappa_A(p)(t) + \overrightarrow{\kappa_A(p)}(t) \\ 0 & 0 \end{pmatrix}\begin{pmatrix} \kappa_A(q_A)(t) \\ 1 \end{pmatrix}.$$

$$\text{(A.5)}$$

We utilize that $\Theta \in SO(3)$ to obtain

$$0 = \frac{d}{dt}I = \frac{d}{dt}\left(\Theta(t)\Theta^{\mathsf{T}}(t)\right) = \dot{\Theta}(t)\Theta^{\mathsf{T}}(t) + \Theta(t)\dot{\Theta}^{\mathsf{T}}(t), \tag{A.6}$$

which implies that $\dot{\Theta}(t)\Theta^{\mathsf{T}}(t) \in \mathbb{R}^{3\times 3}$ is skew symmetric. Every skew symmetric operator on a three-dimensional oriented Euclidean space is the linear map of vector multiplication with a fixed vector, see Arnold [7, Chapter 6.26D, Lemma 2]. As a consequence, there is for every t an $\omega(t)$ such that $\dot{\Theta}(t)\Theta^{\mathsf{T}}(t)x = \omega(t) \times x$ for any $x \in \mathbb{R}^3$. If we define $Q: \mathbb{R}^3 \to T_I SO(3)$ by

$$Q(\omega) = \begin{pmatrix} 0 & -\omega_3 & \omega_2 \\ \omega_3 & 0 & -\omega_1 \\ -\omega_2 & \omega_1 & 0 \end{pmatrix}, \tag{A.7}$$

then

$$\dot{\Theta}(t)\Theta^{\mathsf{T}}(t)x = \omega(t) \times x = Q(\omega(t))x. \tag{A.8}$$

Q is an isomorphism between $\mathbb{R}^3$ and $T_I SO(3)$ and we denote the inverse of Q by Q^{-1}. To assign a physical meaning to $Q(\omega(t))$, we assume that we have pure rotational motion, i.e. that $p(t) = 0$ for $t \geq 0$. Then (A.4) and (A.8) yield

$$\begin{pmatrix} v_A(t) \\ 0 \end{pmatrix} = \begin{pmatrix} \dot{\Theta}(t)\Theta^{\mathsf{T}}(t) & 0 \\ 0 & 0 \end{pmatrix}\begin{pmatrix} \kappa_A(q_A)(t) \\ 1 \end{pmatrix} = \begin{pmatrix} Q(\omega(t)) & 0 \\ 0 & 0 \end{pmatrix}\begin{pmatrix} \kappa_A(q_A)(t) \\ 1 \end{pmatrix}, \tag{A.9}$$

which implies that $\omega(t)$ is the *instantaneous spatial angular velocity*. Furthermore, we define the *instantaneous spatial velocity* $v(t)$ by

$$v(t) = -\dot{\Theta}(t)\Theta^{\mathsf{T}}(t)\kappa_A(p)(t) + \overrightarrow{\kappa_A(p)}(t). \tag{A.10}$$

Then (A.4) implies

$$\begin{pmatrix} v_A(t) \\ 0 \end{pmatrix} = \begin{pmatrix} Q(\omega(t)) & v(t) \\ 0 & 0 \end{pmatrix}\begin{pmatrix} \kappa_A(q_A)(t) \\ 1 \end{pmatrix}. \tag{A.11}$$

In other words, the instantaneous spatial angular velocity $\omega(t)$ and the instantaneous spatial velocity $v(t)$ at a timepoint t determine the speed $v_A(t)$ of a point q with position $\kappa_A(q_A)(t)$ at timepoint t.

The definition of the instantaneous spatial angular velocity ω and the instantaneous spatial velocity v give the differential equations for $(\kappa_A(p)(t), \Theta(t))$ depending on $Q(\omega(t))$ and $v(t)$. To simplify the notation, we write $b(t)$ for $\kappa_A(p)(t)$. Then $(b(t), \Theta(t))$ are the solutions of

$$\begin{aligned} \dot{b} &= Q(\omega)b + v \\ \dot{\Theta} &= Q(\omega)\Theta. \end{aligned} \tag{A.12}$$

The equations (A.12) describe the rigid body kinematics in spatial coordinates.

The kinetic part of the motion for the rigid body is described by the Newton-Euler equations. In case the rigid body is a system of mass points, the Newton-Euler equations are direct consequences of the law of conservation of momentum and the law of conservation of angular momentum. If the rigid body is a system with continuous mass distribution, we obtain the Newton-Euler equations from the generalized postulates by Leonhard Euler, for details see Scheck [118]. Both cases give the same translational equations of motion, expressed in terms of the momentum of the rigid body. The momentum of a rigid body is defined by

$$p = mv \qquad (A.13)$$

where v is the instantaneous spatial velocity of the reference point for the external forces and m is the mass of the rigid body. Because of (A.5), the instantaneous spatial velocity of the center of mass is given by $\dot{b}$, hence the momentum of the rigid body with respect to the center of mass is $p = m\dot{b}$. Let F denote the external forces applied at the center of mass, then the translational equations of motion with respect to the center of mass in spatial coordinates are given by

$$\widehat{m\dot{b}} = F. \qquad (A.14)$$

In other words, the translational velocity of the center of mass is determined by the external forces.

Let M denote the external moments applied to the rigid body, then the rotational equations of motion with respect to the center of mass in spatial coordinates are given by

$$\widehat{J\omega} = M, \qquad (A.15)$$

where the inertia tensor J is given by

$$J = \int_V \rho(q)\,(Q(q))^2\,dV \qquad (A.16)$$

In other words, the change in the angular momentum is determined by the external moments. Overall, the equations of motion of the rigid body in spatial coordinates are given by

$$\begin{aligned}
\dot{b} &= v \\
m\dot{v} &= F \\
\dot{\Theta} &= Q(\omega)\,\Theta \\
\widehat{J\omega} &= M.
\end{aligned} \qquad (A.17)$$

Bibliography

[1] Acebrón, J. A., Bonilla, L. L., Pérez Vicente, C. J., Ritort, F., Spigler, R., 2005. The Kuramoto model: A simple paradigm for synchronization phenomena. Rev. Mod. Phys. 77 (1), 137–185.

[2] Aeyels, D., Rogge, J., 2004. Existence of partial entrainment and stability of phase locking behavior of coupled oscillators. Prog. Theo. Phys. 112 (6), 921–942.

[3] Agoston, M. K., 2005. Computer Graphics and Geometric Modeling — Mathematics. Springer.

[4] Amann, H., Escher, J., 1999. Analysis II. Birkhäuser, translated from German by Silvio Levy and Matthew Cargo.

[5] Appleton, E. V., 1922. The automatic synchronization of triode oscillators. Proceedings of the Cambridge Philosophical Society, 231–248.

[6] Arenas, A., Díaz-Guilera, A., Kurths, J., Moreno, Y., Zhou, C., 2008. Synchronization in complex networks. Physics Reports 469 (3), 93–153.

[7] Arnold, V. I., 1989. Mathematical Methods of Classical Mechanics. Vol. 60 of Graduate Texts in Mathematics. Springer.

[8] Arnold, V. I., 1991. Ordinary Differential Equations. Springer.

[9] Arsie, A., Ebenbauer, C., 2010. Locating omega-limit sets using height functions. J. of Diff. Eqn. 248 (10), 2458–2469.

[10] Artin, M., 1991. Algebra. Prentice Hall.

[11] Atassi, A., Khalil, H., 2000. Separation results for the stabilization of nonlinear systems using different high-gain observer designs. Systems & Control Letters 39 (3), 183–191.

[12] Aulbach, B., 1984. Continuous and Discrete Dynamics near Manifolds of Equilibria. Vol. 1058 of Lecture Notes in Mathematics. Springer.

[13] Austin, D. M., Braam, P. J., 1995. Morse-Bott theory and equivariant cohomology. In: Hofer, H., Taubes, C. H., Weinstein, A., Eduard, Z. (Eds.), The Floer Memorial Volume. Vol. 133 of Progress in Mathematics. Birkhäuser, pp. 123–184.

[14] Bernstein, D. S., 2009. Matrix mathematics: theory, facts, and formulas. Princeton University Press, 2nd edition.

[15] Bhat, S. P., Bernstein, D. S., 2000. A topological obstruction to continuous global stabilization of rotational motion and the unwinding phenomenon. Systems & Control Letters 39 (1), 63–70.

[16] Bhatia, N. P., Szegö, G. P., 1970. Stability Theory of Dynamical Systems. Classics in Mathematics. Springer, originally published as Vol. 161 of Die Grundlehren der mathematischen Wissenschaften in Einzeldarstellungen.

[17] Biagiotti, L., Melchiorri, C., 2008. Trajectory Planning for Automatic Machines and Robots. Springer-Verlag Berlin Heidelberg.

[18] Blekhman, I. I., 1988. Synchronization in Science and Technology. American Society of Mechanical Engineers.

[19] Boccaletti, S., Latora, V., Moreno, M., Chavez, M., Hwang, D.-U., 2006. Complex networks: Structure and dynamics. Physics Reports 424 (4-5), 175–308.

[20] Boyd, S., El Ghaoui, L., Feron, E., Balakrishnan, V., 1994. Linear Matrix Inequalities in System and Control Theory. Vol. 15 of SIAM Studies in Applied Mathematics. Society for Industrial and Applied Mathematics.

[21] Brockett, R. W., 1970. Finite Dimensional Linear Systems. John Wiley and Sons, Inc.

[22] Brockett, R. W., 2001. New issues in the mathematics of control. In: Engquist, B., Schmid, W. (Eds.), Mathematics Unlimited: 2001 and beyond. Springer, pp. 189–220.

[23] Byrnes, C. I., Isidori, A., 1998. Limit sets, zero dynamics, and internal models in the problem of nonlinear output regulation. IEEE Transactions on Automatic Control 48 (10), 1712–1723.

[24] Byrnes, C. I., Isidori, A., 2000. Output regulation for nonlinear systems: an overview. Int. J. of Robust and Nonlinear Control 10 (5), 323–337.

[25] Carmo, M. P. d., 1976. Differential Geometry of Curves and Surfaces. Prentice Hall.

[26] Chaturvedi, N. A., Sanyal, A. K., McClamroch, N. H., 2011. Rigid-body attitude control. IEEE Control Systems Magazine 31 (3), 30–51.

[27] Chen, Z., Huang, J., 2009. Attitude tracking and disturbance rejection of rigid spacecraft by adaptive control. IEEE Transactions on Automatic Control 54 (3), 600–605.

[28] Chicone, C., 2006. Ordinary Differential Equations with Applications. No. 34 in Texts in Applied Mathematics. Springer, 2nd ed.

[29] Choi, M. Y., Kim, H. J., Kim, D., Hong, H., 2000. Synchronization in a system of globally coupled oscillators with time delay. Phys. Rev. E 61 (1), 371–381.

[30] Chopra, N., Spong, M. W., 2009. On exponential synchronization of Kuramoto oscillators. IEEE Transactions on Automatic Control 54 (2), 353–357.

[31] Coddington, E. A., Levinson, N., 1955. Theory of Ordinary Differential Equations. International Series in Pure and Applied Mathematics. McGraw-Hill.

[32] Cox, N., Marconi, L., Teel, A. R., 2012. Hybrid internal models for robust spline tracking. In: Decision and Control (CDC), 2012 IEEE 51st Annual Conference on. pp. 4877–4882.

[33] Crouch, P. E., 1984. Spacecraft attitude control and stabilization: Applications of geometric control theory to rigid body models. Transactions on Automatic Control 29 (4), 321–331.

[34] DeLellis, P., di Bernardo, M., Russo, G., 2011. On quad, Lipschitz, and contracting vector fields for consensus and synchronization of networks. IEEE Transactions on Circuits and Systems I: Regular Papers 58 (3), 576–583.

[35] Dörfler, F., Bullo, F., 2011. On the critical coupling for Kuramoto oscillators. SIAM Journal on Applied Dynamical Systems 10 (3), 1070–1099.

[36] Dörfler, F., Chertkov, M., Bullo, F., 2013. Synchronization in complex oscillator networks and smart grids. Proceedings of the National Academy of Sciences.

[37] Dumortier, F., 2006. Asymptotic phase and invariant foliations near periodic orbits. Proceedings of the American Mathematical Society 134 (10), 2989–2996.

[38] Earl, M. G., Strogatz, S. H., 2003. Synchronization in oscillator networks with delayed coupling: A stability criterion. Phys. Rev. E 67 (3), 036204.

[39] Ebenbauer, C., Raff, T., Allgower, F., 2007. Certainty-equivalence feedback design with polynomial-type feedbacks which guarantee iss. IEEE Transactions on Automatic Control 52 (4), 716–720.

[40] Ermentrout, G. B., Kopell, N., 1991. Multiple pulse interactions and averaging in systems of coupled neural oscillators. Journal of Mathematical Biology 29 (3), 195–217.

[41] Farkas, M., 1994. Periodic Motions. Vol. 104 of Applied Mathematical Sciences. Springer.

[42] Fernandes Vasconcelos, J., Rantzer, A., Silvestre, C., Oliveira, P., 2011. Combination of lyapunov and density functions for stability of rotational motion. IEEE Transactions on Automatic Control 56 (11), 2599–2607.

[43] Franci, A., Chaillet, A., Pasillas-Lépine, W., 2011. Existence and robustness of phase-locking in coupled Kuramoto oscillators under mean-field feedback. Automatica 47 (6), 1193 – 1202.

[44] Francis, B. A., Wonham, M., 1975. The internal model principle for linear multivariable regulators. Applied Mathematics & Optimization 2 (2), 486–505.

[45] Frankel, T., 1965. Critical submanifolds of the classical groups and Stiefel manifolds. In: Cairns, S. S. (Ed.), Differential and Combinatorial Topology — A Symposium in Honor of Marston Morse. Princeton University Press, pp. 37–54.

[46] Freeman, R. A., 2013. A global attractor consisting of exponentially unstable equilibria. In: Proceedings of the American Control Conference (ACC), 2013. pp. 4862–4867.

[47]   Fujisaka, H., Yamada, T., 1983. Stability theory of synchronized motion in coupled-oscillator systems. Progress of Theoretical Physics 69 (1), 32–47.

[48]   Godsil, C., Royle, G., 2001. Algebraic Graph Theory. No. 207 in Graduate Texts in Mathematics. Springer.

[49]   Goldstein, H., Poole, C. P., Safko, J. L., 2000. Classical Mechnics. Addison Wesley, third edition.

[50]   Guckenheimer, J., 1975. Isochrons and phaseless sets. Journal of Mathematical Biology 1 (3), 259–273.

[51]   Guillemin, V., Pollack, A., 1974. Differential Topology. Prentice-Hall, Inc.

[52]   Haddock, J. R., Terjeki, J., 1983. Liapunov-Razumikhin functions and an invariance-principle for functional-differential equations. J. of Differential Equations 48 (1), 95–122.

[53]   Hale, J. K., 1980. Ordinary Differential Equations. Dover, reprint of the work published by Robert E. Krieger Publishing Company.

[54]   Hale, J. K., 1997. Diffusive coupling, dissipation, and synchronization. Journal of Dynamics and Differential Equations 9 (1), 1–52.

[55]   Hale, J. K., Lunel, S. M. V., 1993. Introduction to Functional Differential Equations. No. 99 in Applied Mathematical Sciences. Springer.

[56]   Hartman, P., 2002. Ordinary Differential Equations. Society for Industrial and Applied Mathematics, previously published 2nd ed. Boston: Birkhäuser, 1982. Originally published: Baltimore, Md., 1973.

[57]   Hauser, J., 1994. Converse Lyapunov functions for exponentially stable periodic orbits. Systems & Control Letters 23, 27–34.

[58]   Helmke, U., Moore, J. B., 1994. Optimization and Dynamical Systems. Springer.

[59]   Hoppensteadt, F. C., Izhikevich, E. M., 1997. Weakly Connected Neural Networks. Vol. 126 of Applied Mathematical Sciences. Springer.

[60]   Huang, J., 2004. Nonlinear Output Regulation. SIAM.

[61]   Huygens, C., 1893. Oeuvres Complètes de Christiaan Huygens — Tome Cinquième — Correspondance 1664–1665. La Haye, Martinus Nijhoff, publiées par al Société Hollandaise des Sciences.

[62]   Iggidr, A., Sallet, G., 2003. On the stability of nonautonomous systems. Automatica 39 (1), 167–171.

[63]   Ijspeert, A. J., 2008. Central pattern generators for locomotion control in animals and robots: A review. Neural Networks 21 (4), 642–653.

[64]   Isidori, A., 1995. Nonlinear Control Systems. Springer.

[65] Isidori, A., Byrnes, C. I., 1990. Output regulation of nonlinear systems. IEEE Transactions on Automatic Control 35 (2), 131–140.

[66] Isidori, A., Byrnes, C. I., 2008. Steady-state behaviors in nonlinear systems with an application to robust disturbance rejection. Annual Reviews in Control 32, 1–16.

[67] Isidori, A., Marconi, L., 2011. Nonlinear output regulation. In: Levine, W. S. (Ed.), The Control Systems Handbook — Advanced Methods. Taylor & Francis, pp. 48–1–48–17, section VIII-48.

[68] Isidori, A., Marconi, L., Serrani, A., 2003. Robust autonomous guidance: an internal model approach. Springer Verlag.

[69] Izhikevich, E. M., 1998. Phase models with explicit time delays. Phys. Rev. E 58 (1), 905–908.

[70] Jadbabaie, A., Motee, N., Barahona, M., 2004. On the stability of the Kuramoto model of coupled nonlinear oscillators. In: Proceedings of the American Control Conference (ACC), 2004. pp. 4296–4301.

[71] Jurdjevic, V., 1997. Geometric Control Theory. Cambridge University Press.

[72] Kailath, T., 1980. Linear Systems. Information and System Sciences Series. Prentice-Hall, Inc.

[73] Koditschek, D., 1989. The application of total energy as a Lyapunov function for mechanical control systems. In: Marsden, J. E., Krishnaprasad, P., Simo, J. C. (Eds.), Dynamics and Control of Multibody Systems. Vol. 97 of Contemporary Mathematics. American Mathematical Society, pp. 131–157.

[74] Kokotovic, P. V., 1992. The joy of feedback: nonlinear and adaptive. Control Systems, IEEE 12 (3), 7–17.

[75] Krstić, M., Kanellakopoulos, I., Kokotović, P., 1995. Nonlinear and Adaptive Control Design. Adaptive and Learning Systems for Signal Processing, Communications, and Control. Wiley.

[76] Kuramoto, Y., Nishikawa, I., 1987. Statistical macrodynamics of large dynamical systems. case of a phase transition in oscillator communities. J. of Statistical Physics 49 (3), 569–605.

[77] Lageman, C., Trumpf, J., Mahony, R., 2010. Gradient-like observers for invariant dynamics on a Lie group. IEEE Transactions on Automatic Control 55 (2), 367–377.

[78] Lanczos, C., 1949. The Variational Principles of Mechanics. Oxford University Press.

[79] Landau, L. D., Lifshitz, E., 1976. Mechanics. Butterworth.

[80] LaSalle, J. P., 1976. The Stability of Dynamical Systems. No. 25 in CMBS-NSF Regional Conference Series in Applied Mathematics. SIAM, appendix — Limiting Equations and Stability of Nonautonomous Ordinary Differential Equations by Zvi Artstein.

[81] Lee, J. M., 2006. Introduction to Smooth Manifolds. Vol. 218 of Graduate Texts in Mathematics. Springer.

[82] Lee, W. S., Ott, E., Antonsen, T. M., 2009. Large coupled oscillator systems with heterogeneous interaction delays. Phys. Rev. Lett. 103 (4), 044101.

[83] Li, Z., Chen, G., 2006. Global synchronization and asymptotic stability of complex dynamical networks. IEEE Transactions on Circuits and Systems II: Express Briefs 53 (1), 28–33.

[84] Liberzon, D., Morse, A. S., Sontag, E. D., 2002. Output-input stability and minimumphase nonlinear systems. IEEE Transactions on Automatic Control 47 (3), 422–436.

[85] Lin, Z., Francis, B., Maggiore, M., 2007. State agreement for continuous-time coupled nonlinear systems. SIAM Journal on Control and Optimization 46 (1), 288–307.

[86] Luenberger, D. G., 1964. Observing the state of a linear system. Transactions on Military Electronics 8 (2), 74–80.

[87] Luzyanina, T. B., 1995. Synchronization in an oscillator neural network model with time-delayed coupling. Network: Computation in Neural Systems 6 (1), 43–59.

[88] Maithripala, D. H. S., P., D. W., Berg, J. M., 2005. Intrinsic observer-based stabilization for simple mechanical systems on Lie groups. Siam J. Contr. Opt. 44 (5), 1691–1711.

[89] Marino, R., Tomei, P., 2011. An adaptive learning regulator for uncertain minimum phase systems with undermodeled unknown exosystems. Automatica 47, 739–747.

[90] Marsden, J., Ratiu, T. S., 1999. Introduction to Mechanics and Symmerty. Vol. 17 of Texts in Applied Mathematics. Springer, 2nd Edition.

[91] Mayhew, C. G., Teel, A. R., 2013. Global stabilization of spherical orientation by synergistic hybrid feedback with application to reduced-attitude tracking for rigid bodies. Automatica 49 (7), 1945–1957.

[92] Mazenc, F., Praly, L., Dayawansa, W. P., 1994. Global stabilization by output feedback: examples and counterexamples. Systems & Control Letters 23 (2), 119–125.

[93] Mazur, S., Ulam, S., 1932. Sur les transformations isométriques d'espaces vectoriels, normés. Comptes rendus hebdomadaires des séances de l'Académie des Sciences 194, 946–948.

[94] McDuff, D., Salamon, D., 1998. Introduction to Symplectic Topology. Clarendon Press.

[95] Meyer, G., 1971. Design and global analysis of spacecraft attitude control systems. Nasa tech. rep. 361, Nasa Ames Research Center.

[96] Michalowsky, S., 2012. Attitude control of a multibody spacecraft - a global output regulation approach. Diploma thesis at the Institute for Systems Theory and Automatic Control, University of Stuttgart.

[97] Milnor, J., 1997. Topology from the differentiable viewpoint. Princeton Landmarks in Mathematics. Princeton University Press, originally published by University Press of Virginia, 1965.

[98] Minorsky, N., 1974. Nonlinear Oscillations. Krieger Pub. Co.

[99] Moreau, L., 2004. Stability of continuous-time distributed consensus algorithms. In: Proceedings of the 43th IEEE CDC. Vol. 4. pp. 3998–4003.

[100] Muldowney, J. S., 1990. Compound matrices and ordinary differential-equations. Rocky Mountain Journal of Mathematics 20 (4), 857–872.

[101] Münz, U., Papachristodoulou, A., Allgöwer, F., 2009. Consensus reaching in multi-agent packet-switched networks with nonlinear coupling. Int. J. of Control 82 (5), 953–969.

[102] Murray, R. M., Sastry, S. S., Li, Z. S., 1994. A Mathematical Introduction To Robotic Manipulation. CRC Press.

[103] Nakamura, Y., Tominaga, F., Munakata, T., 1994. Clustering behavior of time-delayed nearest-neighbor coupled oscillators. Phys. Rev. E 49 (6), 4849–4856.

[104] Nijmeijer, H., Rodriguez-Angeles, A., 2003. Synchronization of Mechanical Systems. Vol. 46 of World Scientific Series on Nonlinear Science. World Scientific.

[105] Olfati-Saber, R., Fax, J. A., Murray, R. M., 2007. Consensus and cooperation in networked multi-agent systems. Proceedings of the IEEE 95 (1), 215–233.

[106] Papachristodoulou, A., Jadbabaie, A., 2005. Synchronization in oscillator networks: Switching topologies and non-homogeneous delays. In: Proceedings of the 44th IEEE CDC and European Control Conference. pp. 5692–5697.

[107] Papachristodoulou, A., Jadbabaie, A., Münz, U., 2010. Effects of delay in multi-agent consensus and oscillator synchronization. IEEE Transactions on Automatic Control 55 (6), 1471–1477.

[108] Parrilo, P., 2000. Structured semidefinite programs and semialgebraic geometry methods in robustness and optimization. Ph.D. thesis, California Insitute of Technology.

[109] Pavlov, A., van de Wouw, N., Nijmeijer, H., 2006. Uniform Output Regulation of Nonlinear Systems. Birkäuser.

[110] Pecora, L. M., Carroll, T. L., 1998. Master stability functions for synchronized coupled systems. Phys. Rev. Lett. 80 (10), 2109–2112.

[111] Pikovsky, A., Rosenblum, M., Kurths, J., 2001. Synchronization — A Universal Concept in Nonlinear Sciences. Vol. 12 of Cambridge Nonlinear Science Series. Cam. Univ. Press.

[112] Praly, L., Arcak, M., 2002. On certainty-equivalence design of nonlinear observer-based controllers. In: Decision and Control, 2002, Proceedings of the 41st IEEE Conference on. Vol. 2. pp. 1485–1490.

[113] Roy, R., Thornburg, K. S., 1994. Experimental synchronization of chaotic lasers. Phys. Rev. Lett. 72 (13), 2009–2012.

[114] Rudin, W., 1964. Principles of mathematical analysis. McGraw-Hill.

[115] Salcudean, S., 1991. A globally convergent angular velocity observer for rigid body motion. IEEE Transactions on Automatic Control 36 (12), 1493–1497.

[116] Sanyal, A., Fosbury, A., Chaturvedi, N., Bernstein, D. S., 2009. Inertia-free spacecraft attitude tracking with disturbance refection and almost global stabilization. Journal of Guidance, Control, and Dynamics 32 (4), 1167–1178.

[117] Sarlette, A., Sepulchre, R., 2011. Synchronization on the circle. In: Dubbeldam, J., Green, K., Lenstra, D. (Eds.), The Complexity of Dynamical Systems: A Multi-disciplinary Perspective. Wiley, pp. 213–240.

[118] Scheck, F., 2007. Theoretische Physik 1 — Mechanik. Springer, achte Auflage, German.

[119] Schmidt, G. S., Ebenbauer, C., Allgöwer, F., 2010. Synchronization conditions for Lyapunov oscillators. In: Decision and Control (CDC), 2010 49th IEEE Conference on. pp. 6230–6235.

[120] Schmidt, G. S., Ebenbauer, C., Allgöwer, F., 2012. A solution for a class of output regulation problems on SO(n). In: Proceedings of the American Control Conference (ACC), 2012. pp. 1773–1779.

[121] Schmidt, G. S., Ebenbauer, C., Allgöwer, F., 2013. On the differential equation $\dot{\Theta} = (\Theta^T - \Theta)\Theta$ with $\Theta \in SO(n)$. http://arxiv.org/abs/1308.6669.

[122] Schmidt, G. S., Ebenbauer, C., Allgöwer, F., 2013. Output regulation for attitude control: a global approach. In: Proceedings of the American Control Conference (ACC), 2013. pp. 5259–5264.

[123] Schmidt, G. S., Ebenbauer, C., Allgöwer, F., 2014. Output regulation for control systems on se(n): A separation principle based approach. Transactions on Automatic ControlAccepted, Tentatively scheduled to appear in December 2014.

[124] Schmidt, G. S., Michalowsky, S., Ebenbauer, C., Allgöwer, F., 2013. Global output regulation for the rotational dynamics of a rigid body. Automatisierungstechnik 61 (8), 567–582.

[125] Schmidt, G. S., Münz, U., Allgöwer, F., 2009. Multi-agent speed consensus via delayed position feedback with application to Kuramoto oscillators. In: Proceedings of the European Control Conference. pp. 2464–2469.

[126] Schmidt, G. S., Papachristodoulou, A., Münz, U., Allgöwer, F., 2012. Frequency synchronization and phase agreement in Kuramoto oscillator networks with delays. Automatica 48 (12), 3008–3017.

[127] Schuster, H. G., Wagner, P., 1989. Mutual entrainment of two limit cycle oscillators with time delayed coupling. Prog. Theo. Phys. 81 (5), 939–945.

[128] Sepulchre, R., Jankovic, M., Kokotović, P., 1997. Constructive Nonlinear Control. Springer.

[129] Sepulchre, R., Paley, D. A., Leonard, N. E., 2007. Stabilization of planar collective motion: All-to-all communication. IEEE Transactions on Automatic Control 52 (5), 811–824.

[130] Sepulchre, R., Paley, D. A., Leonard, N. E., 2008. Stabilization of planar collective motion with limited communication. IEEE Transactions on Automatic Control 53 (3), 706–719.

[131] Serrani, A., Isidori, A., Marconi, L., 2000. Semiglobal robust output regulation of minimum-phase nonlinear systems. International Journal of Robust and Nonlinear Control 10 (5), 379–396.

[132] Serrani, A., Isidori, A., Marconi, L., 2001. Semi-global nonlinear output regulation with adaptive internal model. IEEE Transactions on Automatic Control 46 (8), 1178–1194.

[133] Shuster, M. D., 1993. A survey of attitude represenations. The Journal of the Astronautical Sciences 41 (4).

[134] Siciliano, B., Khatib, O. (Eds.), 2008. Springer Handbook of Robotics. Springer.

[135] Sidi, M. J., 1997. Spacecraft Dynamics and Control. Cambridge University Press.

[136] Slotine, J.-J. E., Wang, W., El Rifai, K., 2004. Contraction analysis of synchronization in networks of nonlinearly coupled oscillators. In: Proceedings of the 16th Int. Symp. Mathematical Theory of Networks and Systems.

[137] Strauss, A., Yorke, J. A., 1967. On asymptotically autonomous differential equations. Mathematical systems theory 1 (2), 175–182.

[138] Strogatz, S. H., 2000. From Kuramoto to Crawford: Exploring the onset of synchronization in populations of coupled oscillators. Physica D 143 (1-4), 1–20.

[139] Strogatz, S. H., 2001. Exploring complex networks. Nature 410 (6825), 268–276.

[140] Strogatz, S. H., 2003. Sync. Hyperion.

[141] Strutt, J. W. B. R., 1877. Theory of Sound. Vol. 1. Macmillan.

[142] Strutt, J. W. B. R., 1878. Theory of Sound. Vol. 2. Macmillan.

[143] Teschl, G., 2011. Ordinary Differential Equations and Dynamical Systems. Vol. 140 of Graduate Studies in Mathematics. American Mathematical Society.

[144] Thews, K., 1989. Der Abbildungsgrad von Vektorfeldern zu stabilen Ruhelagen. Archiv der Mathematik 52 (1), 71–74.

[145] Thieme, H. R., 1992. Convergence results and a Poincaré-Bendixson trichotomy for asymptotically autonomous differential equations. Journal of Mathematical Biology 30, 755–763.

[146] Verwoerd, M., Mason, O., 2008. Global phase-locking in finite populations of phase-coupled oscillators. Journal of Applied Dynamical Systems 7 (1), 134–160.

[147] Wen, J. T.-Y., Kreutz-Delgado, K., 1991. The attitude control problem. IEEE Transactions on Automatic Control 36 (10), 1148–1162.

[148] Wilson, F. Wesley, J., 1969. Smoothing derivatives of functions and applications. Transactions of the American Mathematical Society 139, 413–428.

[149] Winfree, A. T., 2001. The Geometry of Biological Time. Springer.

[150] Wonham, M., 1985. Linear Multivariable Control. Vol. 10 of Stochastic Modeling and Applied Probability. Springer.

[151] Yeung, M. K. S., Strogatz, S. H., 1999. Time delay in the Kuramoto model of coupled oscillators. Phys. Rev. Letters 82 (3), 648–651.

[152] Zubov, V. I., 1964. Methods of A. M. Lyapunov and their Applications. P. Noordhoff.